T0219856

Data Science and Analytics for SMEs

Consulting, Tools, Practical Use Cases

Afolabi Ibukun Tolulope

Apress®

Data Science and Analytics for SMEs: Consulting, Tools, Practical Use Cases

Afolabi Ibukun Tolulope
London, United Kingdom

ISBN-13 (pbk): 978-1-4842-8669-2   ISBN-13 (electronic): 978-1-4842-8670-8
https://doi.org/10.1007/978-1-4842-8670-8

Copyright © 2022 by Afolabi Ibukun Tolulope

Managing Director, Apress Media LLC: Welmoed Spahr
Acquisitions Editor: Shiva Ramachandran
Development Editor: James Markham
Coordinating Editor: Jessica Vakili

Distributed to the book trade worldwide by Springer Science+Business Media New York, 1 New York Plaza, New York, NY 10004. Phone 1-800-SPRINGER, fax (201) 348-4505, e-mail orders-ny@springer-sbm.com, or visit www.springeronline.com. Apress Media, LLC is a California LLC and the sole member (owner) is Springer Science + Business Media Finance Inc (SSBM Finance Inc). SSBM Finance Inc is a **Delaware** corporation.

For information on translations, please e-mail booktranslations@springernature.com; for reprint, paperback, or audio rights, please e-mail bookpermissions@springernature.com.

Apress titles may be purchased in bulk for academic, corporate, or promotional use. eBook versions and licenses are also available for most titles. For more information, reference our Print and eBook Bulk Sales web page at http://www.apress.com/bulk-sales.

Printed on acid-free paper

Table of Contents

About the Author

 **Afolabi Ibukun** is a Data Scientist and is currently an Assistant Professor in Computer Science at Northeastern University London. She holds a B.Sc in Engineering Physics, an M.Sc and Ph.D in Computer Science. Afolabi Ibukun has over 15 years working experience in Computer Science research, teaching and mentoring. Her specific areas of interest are Data & Text Mining, Programming and Business Analytics. She has supervised several undergraduate and postgraduate students and published several articles in international journals and conferences. Afolabi Ibukun is also a Data Science Nigeria Mentor (`https://www.datasciencenigeria.org/mentors/`) and currently runs a Business Analytics Consulting and Training firm named I&F Networks Solutions."

08021247616
ibukunafolabi0909@gmail.com
linkedin.com/in/afolabi-ibukun-051777a6
github.com/ibkAfolabi

About the Technical Reviewer

Hitesh Hinduja is an ardent Artificial Intelligence (AI) and Data Platforms enthusiast currently working as Senior Manager in Data Platforms (Azure) and AI at Microsoft. He worked as a Senior Manager in AI at Ola Electric, where he led a team of 20+ people in the areas of machine learning, statistics, computer vision, deep learning, natural language processing, and reinforcement learning. He has filed 14+ patents in India and the United States and has numerous research publications under his name. Hitesh had been associated in research roles at India's top B-schools – Indian School of Business, Hyderabad, and the Indian Institute of Management, Ahmedabad. He is also actively involved in training and mentoring and has been invited as a guest speaker by various corporates and associations across the globe.

Hitesh is an avid learner and enjoys reading books in his free time.

Acknowledgments

First of all, I would like to acknowledge God almighty for making it possible for me to write this book. I would also like to thank my husband Oluwafemi Afolabi for his support and encouragement that has made this book a reality. I deeply appreciate Prof. Olufunke Oladipupo and Dr. Joke Badejo who have taught me a lot, both as a data scientist and in other aspects of life. Lastly, I would like to say thank you to Timileyin Owoseni and Christabel Uzor, my M.Sc. students who also helped with the book. I will not forget my wonderful students that I have been fortunate to teach and advise, Obinna Okorie, Temi Oyedepo, and many others; I learned a lot from them. I would like to appreciate afrimash.com for the opportunity to learn practical data science consulting.

Preface

This book is written from the perspective of offering a Business Analytics service as a product. It helps to understand how to package your analytics solution as a product that can be offered as a consulting service. Some of these products are customer loyalty, market segmentation, sales and revenue increase, etc. It is also particularly focused on small businesses and their peculiarity in analytics. Understanding the contents of the book will help anyone interested in applying data analytics to make a difference in small businesses achieve such, starting from the beginner's level. It uses a do-it-yourself approach to analytics, and the tools used are easily available online and are nonprogramming based.

The book teaches the tricks and techniques of Business Analytics consulting for small businesses. In particular, readers will be able to create and measure the success of their analytics project. The book also provides a career guide and helps to jump-start the world of Business Analytics consulting career. The approach in the book is to focus on popular problems in the small and medium business world that have data science solutions and then introduce the technique and how to use it to solve the problems. Readers will not only learn the fundamental techniques used in solving these problems, but they will also experience how to use them in practical use cases and problem scenarios. The techniques are taught in a simple way, but the book is supported with a lot of reference and resource material that can help build more mastery on the techniques.

The book is divided into four major parts. Part 1 (Chapters 1 and 2) explains the fundamental concepts explored in this book, such as data science, data science for business (Business Analytics), and what it takes to carry out any analytics project both generally and specifically for a small business. In this part, we also explore issues around data and how to manage and prepare it for the analytics project with practical examples. Part 2 (Chapters 3 and 4) focuses on analytics consulting and explains how to navigate your way through to becoming successful in the data analytics consulting space. It also gives a detail of the phases involved in Business Analytics consulting. Part 3 (Chapters 5–8) is focused on the data mining techniques common with small businesses, and this is expressed in an approach that first explains the basic concepts of these techniques in a simple way and then uses a real small business problem scenario for the practical application. This part is practical oriented and based on case study problems experienced by small businesses. In this book, we will explore five major practical business problem scenarios and several small business problems for illustration. These are covered in Chapters 5–8. The techniques used demonstrate how to solve these problems. It is important to say here that despite using a particular problem as a case study, it is not only in this situation that the approach can be deployed, but it can be used in other similar problem scenarios. The techniques selected are based on their popularity in practice, and they fall under the broad classification of prediction (predicting numerical outcome), classification (predicting categorical outcome), and descriptive analytics. Finally, Part 4 brings the consulting principles to life by using an SME case study to model the already explained consulting phases in Part 2 and adopting the appropriate techniques among the ones explained in Part 3. Although each chapter stands alone, we advise that you read Part 1 before proceeding to Part 3, and Part 2 before proceeding to Part 4.

The tools used for the practical examples are RapidMiner Studio and Gephi. The book is written such that all the RapidMiner Studio and Gephi screenshots are included with details of how to run them. GitHub will also be used to store the practical project.

This book goes beyond explaining the techniques to giving an experience of applying the recommendations from the modeling to get results. In particular, we use a sample of business scenarios experienced in the past for the use cases. The book is also supported with a real-life business group on Telegram (https://t.me/+kSSQjNhhz6p1ZTkO) where we harvest business problems and encourage readers to be a part of solving the problems. We have in this book real-life business case studies (from the Telegram group) that can be used as a reference. The book also comes with links to YouTube videos that help to explain some of the concepts better.

This book, together with the solutions to the exercises and more application scenarios and data science techniques, is available as an online course and training; you can visit datasciencenaija.com for more details.

We appreciate reader feedback. We would like to know what you think about this book, good or bad. To give us general feedback, simply send an email to ibukunafolabi0909@gmail.com. Also, the data and more information about the book can also be obtained at www.datasciencenaija.com/book.

CHAPTER 1

Introduction

In this chapter, we introduce data science generally and narrow it down to data science for business which is also referred to as Business Analytics. We then give a detailed explanation of the process involved in Business Analytics in the form of the Business Analytics journey. In this journey, we explain what it takes from start to finish to carry out an analytics project in the business world, focusing on small business consulting, even though the process is generic to all types of business, small or large. We also give a description of what small business refers to in this book and the peculiarities of navigating an analytics project in such a terrain. To conclude the chapter, we talk about the types of analytics problems that are common to small businesses and the tools available to solve these problems given the budget situation of small businesses when it comes to analytics projects.

1.1 Data Science

In simple terms, data science refers to the ability to take data, generate an understanding from the data, process the data, extract value from the data, visualize the data, and present it in such a way that decisions can be made from this presentation. Data science is described as the process of extracting knowledge from huge amounts of data. It is an intersection of Mathematics, Statistics, Visualization, and Artificial Intelligence. Artificial Intelligence is a superset of machine learning, which is a superset of deep learning.

© Afolabi Ibukun Tolulope 2022
A. I. Tolulope, *Data Science and Analytics for SMEs*,
https://doi.org/10.1007/978-1-4842-8670-8_1

1.2 Data Science for Business

Data science for business is popularly referred to as Business Analytics, which is how we will refer to it in this book. The activity and art of using quantitative data to inform decision-making is known as Business Analytics. Several organizations have different interpretations of the term. Data visualization and reporting are used in Business Analytics to comprehend "what happened and" what is happening (Business Intelligence).[1] The goal of Business Analytics is to assist you to focus on the datasets that will help you increase your company's revenue, productivity, and efficiency.[4]

In Business Analytics, we want to see how we can analyze data from various sources particularly keeping in mind the KPIs (key performance indicators) of the business in real time and use this to support strategic business decisions.

Comparing Business Analytics and data science, we discover that Business Analytics deals with extracting meaningful insights from the visualized data for making decisions, while data science is more directed to taking raw data and using algorithms, statistical models, and computer programming to extract valuable insights. In this book, we combine both the perspective of Business Analytics and data science and refer to it as data science for business.

Business Analytics has major application types that include

1. *Financial analytics*: This can be characterized as data analytics aimed at solving specific business queries and predicting future financial outcomes. Financial analytics aims to shape business strategy using trustworthy, verifiable information rather than intuition. It gives companies extensive views of their financial data as well as the tools they need to understand significant patterns and take action to

enhance their performance (www.techtarget.com/
searcherp/definition/financial-analytics).
Organizations such as McAfee, Deloitte, and Wiley
are seeing reductions in costs, gains in efficiency,
and so on through financial analytics.

2. *People analytics*: People analytics is a data-driven
    approach to workplace management. People are
    the most important assets to organizations, and if
    they are managed better, it will impact the business
    for better. We use it to answer questions like: Who
    should we hire? Who should we promote? What are
    the best career paths for the company? What are
    the patterns of collaboration like? Who are the key
    communicators, and what can we do to increase
    communication? And so forth. Using people
    analytics, companies such as Google, Amazon,
    Cisco, and many others are able to understand how
    exactly to engage, retain, and ensure productivity
    from their people.

3. *Customer analytics*: This is when data is used to
    better understand the customer's composition,
    needs, and satisfaction. Additionally, the enabling
    technology is used to divide customers into groups
    based on their behavior, to identify general trends,
    and to generate customized marketing and sales
    activities.[8] In domains such as banking, insurance,
    or finance, customer analytics can help to
    understand customer life value, attrition, and so on.
    Customer analytics is heavily used in ecommerce
    websites like Amazon, Flipkart, and many more.

4. *Operation analytics*: The term "operational analytics" refers to a sort of Business Analytics that focuses on enhancing current operations.[11] This sort of Business Analytics, like others, entails the use of a variety of data mining and data aggregation technologies in order to obtain more transparent data for business planning.[9] Operational analytics has been used by Loom, Bold Penguin, and Atrium inspiring things such as unlocking collaboration between customer success and sales to reach customers with a unified front and many more.

1.3 Business Analytics Journey

Business Analytics life cycle can be referred to as information-action value chain[2] or Business Analytics journey. It is the process of linking the data and its source to the analysis and the result of the analysis. It also includes communicating the results of the analysis and assessing if the goals of the analytics project have been met.

The information-action value chain (Business Analytics journey) consists of about nine stages as captured in Figure 1-1.

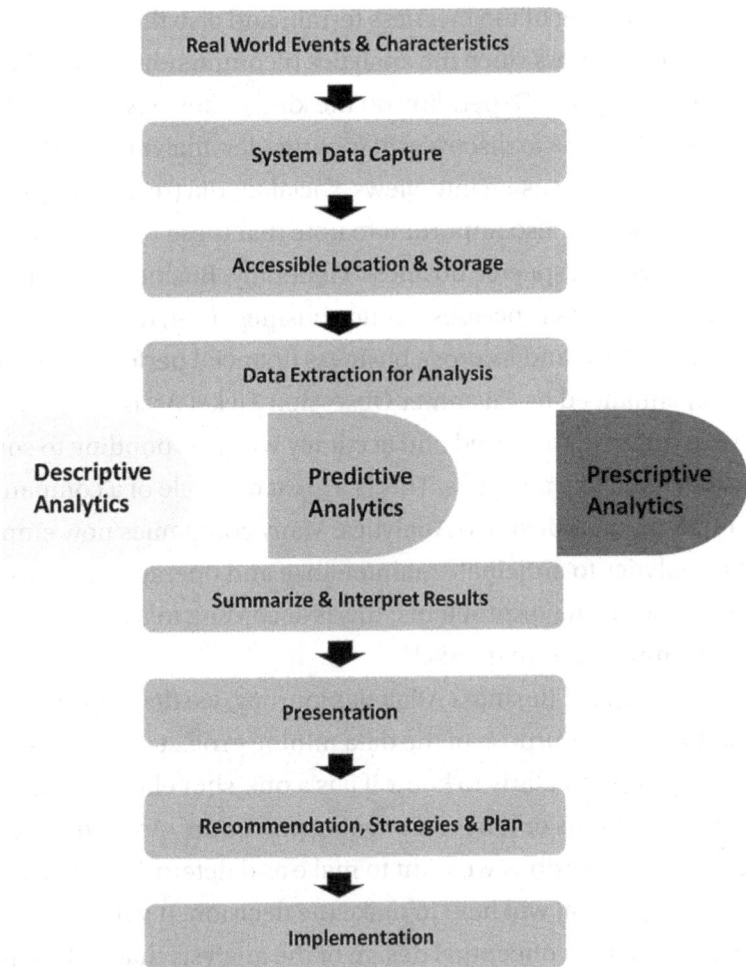

Figure 1-1. *Business Analytics journey*

Events in Real Life and Description

The problems that analytics can solve for business comes majorly by discovering needs of a particular business, particularly the needs that can be solved by analytics. It is important to note that before you can build a successful analytics project that will yield results, you need to have a domain understanding of the business terrain and also the typical success strategies of the business since the analytics recommendation will be used to improve this purpose. Depending on the kind of business in question, there are several means to discover their particular analytics needs. Some of these means include using interviews, social media (text mining), focus groups, and so on. It is also important to note that these needs should be focused on the goal of specific business. Generally, Business Analytics helps to boost business processes, reduce business cost, drive business strategy, and monitor and improve business financial performance. Uber, for example, enhanced its Customer Obsession Ticket Assistant (COTA) in early 2018 to improve the speed and accuracy while responding to support tickets using predictive analytics. This is a great example of a company that has implemented Business Analytics. Many companies now employ predictive analytics to anticipate maintenance and operational concerns before they become major problems; this is according to a KPMG analysis on emerging infrastructure trends.[12]

In this first stage of Business Analytics journey, we develop an understanding of the purpose of the data mining project or Business Analytics project particularly to know if it is a one-shot effort to answer a question or questions or if it is an ongoing procedure. After this, we then highlight the decisions we want to make and determine or suggest the analysis output that will help to make the decision. It will help to have a design that is a conceptual design of the analysis that will create the output.

The conceptual business model is used to understand data in the business domain and how the business actually works. It particularly helps to understand the importance of context in solving analytical problems. It also helps us to understand how important elements relate together. Since Business Analytics solves problems, that is, it answers questions, it is therefore important to know at this stage what questions are worth asking in the course of the analytics project. This conceptual business model will help us to do that.

In this stage of dealing with real-world events, we also want to understand what exactly contributes to the data that we are to be using for the analytics project. For example, the data might come from someone making a purchase. It could also come from using a product or service. As a matter of fact, data could come from anything in the problem domain. Just as is referred to in the definition of Business Analytics, there is a need to focus on the data which is targeted toward contributing to the solution. It is not just about having a large volume of data but focusing on the data needed to solve the problem or questions highlighted. The following scenario gives a real SME business example of interpreting this stage of the analytics journey.

A real business problem: Alegria Recyclers Ltd (www.alegriarecyclers.com.ng/) is a small business which deals in recycling waste materials. Alegria has just two permanent staffs and about ten ad hoc staffs. Alegria deals in recycling hospital wastes like the X-ray films. Their customers are various types of hospitals and clinics. The task at hand for the data scientist is come up with the best actionable strategy for targeting future customers.

The conceptual business model of this business is captured in Figure 1-2. It captures most of the activities of the business. As a Business Analytics consultant, you need to understand the business you are consulting for, by first developing the conceptual model; this will help you to know how to capture the real-world events in the business domain, that is, the data needed and the importance of different types of data. It will also help you to understand how the results of your analysis will be used optimally to get the best results in your analytics project. The first step talks about looking for prospects; this is where you want to explore data of the type of clients that they have worked with in the past, what are the attributes that describe such prospects, what kind of materials are obtained from these different types of clients, and so on; all the information, for example, can help to build clustering models that will help Alegria to understand their prospects and many more. In the second step of the conceptual model which deals with how Alegria contacts the prospects in step one, you want to store information on the time of contact, location, etc. For each of the remaining steps, such as negotiating for collecting of waste, collecting the waste, recycling the waste, and making revenue from the waste, you will have to bring out the data attributes that could be captured. This is how you are able to use the conceptual business model to have a complete understanding of the data in the business domain.

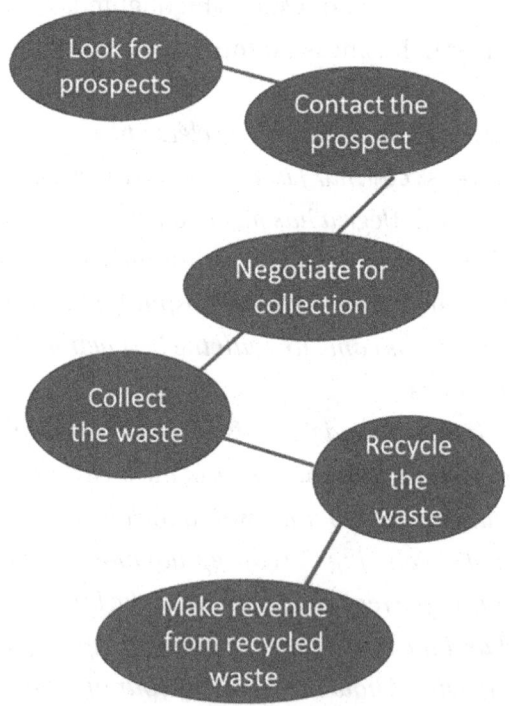

Figure 1-2. *Conceptual business model of Alegria Recyclers Ltd*

Capturing the Data

After successfully identifying what will form the basis of your data and the type of data using the conceptual business models, the next thing is to capture the data and store it in a form that can be used for analytics. Capturing the data deals with the mechanism that captures the physical or digital representation of that real-world phenomenon and puts it somewhere. In this stage, there is a need to identify what is to be stored or retrieved (i.e., if the data has been stored before). Sometimes, it might involve feature selection and representation (which will be explained in Section 2.1). The peculiarity of small businesses when it comes to capturing the data is that there is a lot of data literacy needed to be done due to the fact that most of these businesses are characterized with lack of an enterprise management system that manages their data; also, they particularly don't know what the data needs of their business are. There is therefore a need for data integrity evangelism so as to get the best out of the analytics project. For this reason, the task of capturing data for analysis is a little challenging; most of the data is in the form of excel sheets that might not be particularly analytics ready. There are several data cleaning and preprocessing functions in Microsoft Excel which can be used. Also, there might be a need to use more interactive data visualization tools like Tableau (www.tableau.com/) and Power BI (www.microsoft.com/en-us/download/details.aspx?id=58494) for this purpose of capturing and preparing the data for analysis. There are other tools available for free or for a token on the Internet that can help capture small business processes which will help make available data at the back end that can jump-start the analytics process no matter how small the business is. Google Analytics, for example, can help small businesses, particularly ecommerce businesses, to create actionable strategies from automatically generated data from the ecommerce platform. This data can be used to measure marketing campaign performance, return on investment, and many more (analytics.google.com). For businesses that have been able to put the correct data gathering and governance procedures in place, the data capturing stage is much more easy to navigate as all that is left is to check the appropriateness of the data for the particular analytics goal.

Accessible Location and Storage

In this stage, you try to understand where the data you are working with comes from and then bring them together on one platform. There might be a need to pull data together from different databases which could be internal (e.g., past purchases made by customers) or external (credit ratings, census data, etc.). Most often, this is referred to as a data warehouse or enterprise data warehouse. Currently, there are a variety of emerging technologies to do this both physical and virtual.

There are several sources that data might come from; we will learn more about this in the upcoming Section 2.1. In situations where enterprise management systems or software exist no matter how small, the primary source of data for business is the operational databases or application program as explained in the conceptual business model. There are other sources of data in business, for example, scannable barcodes, point-of-sale (POS) devices, mouse click trails, global positioning satellite (GPS), and so on. The Internet is also a rich source of external data, whether stored in databases or from social media platforms.

Extracting Data for Analysis

In preparation for analysis, a good question to answer is: How many variables and how much data do you need to use for your modeling? According to Galit et al. (2019, p. 29), we can have ten records for every predictor variable, and Delmaster and Hancock (2001, p. 68) also recommend that for classification procedures, we can have at least $6 \times m \times p$ records, where m is the number of outcome classes and p is the number of variables.

In extracting data for analysis, it's important to sample data; this is important because it helps to get a good sample proportion out of the population since we have some algorithms that cannot handle large amounts of data for statistical analysis. Hence, representing a good sample proportion out of our population will be a good practice to perform analysis and come up with several hypotheses.

Also while preparing your data for analysis, you would want to explore, clean, and preprocess the data, verify that the data are in reasonable condition, handle missing data, and check if the values of attributes are in reasonable range, given what you would expect from each variable. At this stage, you also want to check for outliers, normalize your data if necessary, and ensure consistency in the definitions of fields, units of measurement, time periods, and so on. Finally, at this stage, you want to handle categorical variables, converting to dummies where necessary.

Another major step in preparing data for analysis is data visualization which will be explained in detail in Section 2.5. Visualization can help to take care of most of the data preparation steps mentioned earlier and is also used for variable derivation and selection. Visualization can help to determine the appropriate bin sizes in case binning of numerical variables will be needed and combine categories as part of the data reduction process. In case the cost of data collection is high, data visualization can be used to determine which variables and metrics are actually useful. The visualization also helps us to understand the data structure and clean the data to remove unexpected gaps or "illegal" values. Visualizations will also help to identify outliers, discover initial patterns, and come up with questions that can be explored in the course of the analysis.

In extracting data for analysis, data reduction helps to eliminate variables that are not needed. In addition to this, feature extraction methods are used to reduce the data's dimension and identify the majority of the components that represent the data. Data partitioning (in Chapter 6) and transformation are very important in preparing our data for analytics. Data transformation is carried out in situations where you need to turn variables like amount earned into a range between N50,000 and N100,000. There are also situations where you need to create new variables from what you have, for example, creating a variable true or false capturing if a customer showed up or not from the purchase data on customers. It is also important for us to know what each variable means and whether to include it in the analysis or not.

Data Analytics

In this stage, we determine the data mining task to be carried out. This ranges from descriptive analytics, diagnostic analytics, predictive analytics, prescriptive analytics, and cognitive analytics. Sometimes, you might need to combine these tasks or use them as a prerequisite to each other. It is important to note here that one method isn't more important than another; it all depends on the end goal or the purpose of the Business Analytics project.

Descriptive analytics helps to describe what things look like now or what happened in the past. In descriptive analytics, we use available information to better understand the business environment and apply the knowledge along with business acumen to make better decisions. Descriptive analytics makes use of simple aggregations, cross-tabulations, and simple statistics like means, medians, standard deviations, and distributions, for example, histograms. Some advanced descriptive analytics includes associations or clustering algorithms. An example of the question that descriptive analytics helps to answer is: *What type of customers are renting our equipment?* Sometimes, we use descriptive analytics to identify the link between two variables; this field of study is known as association rule mining. A supermarket, for example, might collect information on customer purchase behavior. The supermarket can use this information to discover which products are usually purchased together and sell them accordingly. This is known as market basket analysis. In addition, clustering can be used to identify groupings and structures in data that are "similar" in some sense without relying on existing data structures.

Predictive analytics takes what we know of what happened in the past to predict what will happen in the future. It makes use of advanced statistics such as regression algorithms that include linear, logistic, and tree-based algorithms, neural networks, and so on. Classification and regression tasks fall under predictive analytics. Classification deals with generalizing known structures to apply to new data. For example, an email program might attempt to classify an email as "legitimate" or as "spam." Regression attempts to determine the strength and character of the relationship between one dependent variable and a series of other variables. An example of the question that predictive analytics helps to answer is: *Who rented our equipment in the past, predict who will rent in the future?* Predictive analytics is typically an iterative process in which we try multiple variants of the same algorithm, that is, choosing different variables or settings within the algorithm. The choice of what to settle for is usually decided by feedback from the algorithm's performance on validation data and how well it is performing on real-world data. It is important to note here that machine learning algorithms are data dependent, so this iterative process must be observed for each case study data before concluding which algorithm to settle for.

Prescriptive analytics links analysis to decision making by providing recommendations on what we should do. It helps us to know which customers to target and what choice we should make and usually involves the integration of numerical optimization techniques with business rules and financial models.

There are other types of tasks which include the diagnostic analytics and the cognitive analytics.

Diagnostic analytics takes past performance and uses it to determine which elements influence specific trends. It helps to answer the question *"why did it happen?"*, thus creating a clear connection between data and behavioral patterns. It enables data analysts to drill into the analytics to uncover patterns, trends, and correlations, and it uses more complicated queries. Regression analysis, anomaly detection, clustering analysis, and other techniques are among those used.[5]

Cognitive analytics: The descriptive, diagnostic, predictive, and prescriptive analytics are gradually evolving currently to include specialized areas such as cognitive analytics, automated analytics, smart analytics, and more.[5] Artificial Intelligence and Machine Learning techniques are combined with data analytics approaches in cognitive analytics. In the field of Business Analytics, cognitive analytics solutions assist companies in making key business decisions and reaching conclusions based on existing knowledge bases.[5]

The choice of which task or algorithm to use or how to combine them is determined by the size of the dataset; the types of patterns that exist in the data, that is, whether the data meets some underlying assumptions of the method; how noisy the data is; and, most importantly, the particular goal of the analysis. The approach used in data mining is to apply several different methods and select the one that is most useful for the goal at hand. For a more comprehensive understanding of how to evaluate models, check Galit Shmueli et al. (2018, p. 117).

Summarize and Interpret Results

In this stage of the Business Analytics journey, we interpret and summarize the results obtained from the data analytics stage. It is very important to factor in the context of the business. There is a need for a clear understanding of how the business works so you can understand what the results are telling you. We interpret the results of analysis using charts, graphs, statistical tables, and so on in order to make it easy to see what is going on. In doing this, simplicity is the key; we need to narrow all that has been done to a few key points and determine whether you have successfully answered the questions we set out to answer. In Business Analytics, when interpreting the results, you must not forget to state (by linking the interpreted results with the problem scenario) what the company stands to gain from this analysis.

Presentation

In presenting the results, you need a clear understanding of how the business works; then with the help of charts, graphs, or tables, you will be able to present what is going on. You will also need to supplement the figure with a short narrative explaining what they mean. Simplicity is the word here, regardless of how complex the analytics procedure is; at this point, you will need to reduce the details and sieve out all your analysis into a few key points. At this point, you want to check if you have been able to successfully answer all the questions you set out to answer at the beginning of the analytics project. It's interesting to know that sometimes you might even discover that you were asking or pursuing the wrong goal or question to begin with. Do not forget that as a business analyst, solid presentation skills are required to be able to communicate your analytics results efficiently. Depending on the tools you are using to make your presentation (Tableau dashboard, Microsoft PowerPoint, etc.), you will need to develop mastery in its use and ability to use it to communicate effectively.

Recommendations, Strategies, and Plan

After carrying out the analysis, there is a need to deploy the model by integrating the model into operational systems and running it on real records to produce decisions or actions. For example, the model might be applied to a data on the existing customers of an ecommerce business, and the action might be "a campaign targeted toward customers most likely to churn."

In making these recommendations and providing strategic plans, there will be a need to back up your recommendations with executive summary of the analytics process and focus on the business value of the proposed plan. There is also a need to constantly link the plan to the return on investment, particularly considering the context of the analytics project. In this phase, use the insights obtained from previous steps to create a plan for taking some action, from high-level strategy to specific actions to be taken the strategy will involve you creating several alternative course of action from your analysis and the advantages and disadvantages of each so that the stakeholders can determine with one to opt for.

Implementation

In the implementation stage, the result of the analysis is put to test live. The nature of this implementation will be determined by your strategy in the previous step. In this stage, you will have to monitor the success metrics you highlighted related to each goal in the first step of the analytics journey. Depending on whether the results are successful or not, you might need to iterate the process based on lessons learned in the analytics journey. This entire process is iterative in nature. Also, after implementation, there is a need to use feedback from the users to iterate the analytics process for better analytics results and interpretation of results. The stage to be repeated in this iteration is determined by the nature of the feedback. There are situations where there will be a need to revisit the system data capture stage or even the data analysis stage.

1.4 Small and Medium Enterprises (SME)

According to Uyi Akpata in a report by PwC Nigeria,[3] the MSME (micro, small, and medium enterprises) sector is an economy's growth engine, contributing to its development, job generation, and export, among other things. According to the World Bank, MSMEs account for over 90% of all enterprises and more than 50% of all jobs worldwide. In emerging economies, formal SMEs can account for up to 40% of national income (GDP).

According to the Bank of Industry (BOI) in a report by PwC[3], SMEs can be defined using Table 1-1.

Table 1-1. *SME's Definition*

Enterprise Category Indicator	Micro Enterprise	Small Enterprise	Medium Enterprise
Number of employees	<=10	>11<=50	>51<=200
Total assets (N)	<=5million	>5<= 100million	>100<=500million
Annual turnover (N)	<=20million	<=100million	<=500million

In the same report, the most pressing problem currently faced by SMEs is obtaining finance, followed by finding customers. For SMEs, the task of finding customers is mostly captured by marketing analytics, analytics in customer relationship management, and so on, which takes the lion share of Business Analytics.

1.5 Business Analytics in Small Business

So far, some SMEs have benefited from data science, the likes of which include Goodvine Group[8] and so on, but there is still a lot of ground to cover. SMEs can equally benefit from data science by solving the same challenges as large businesses which include customer acquisition cost, churn, sales forecasting, logistics, or capturing market share; the only difference is that this has to be done with fewer resources. Although small businesses may not be capturing a larger volume of data as large enterprises, the variety and velocity are often the same. The good thing is that, so far, they are able to use the data quickly and efficiently and can compete with larger competitors in the same space. Interestingly, one of the edges that small businesses have over larger ones when it comes to maximizing data science is that they do not need a large data science team to get value out of data.

The major problems of small and medium businesses when it comes to applying data science to improve revenue include the following:

The amount of data collected: Due to technology advancement, there is an enormous amount of data available in small businesses, but these businesses lack a data system that efficiently collects and organizes information.

Data integration: For analysis to be complete and accurate, there is a need to bring together data across multiple, disjointed sources; currently, this is done manually and can be time-consuming and cumbersome.

Data that lacks quality and integrity: Garbage in, garbage out. There are no accurate insights that we can get from data that is full of errors and does not reflect the whole problem scenario, sometimes referred to as asymmetrical data. For more details of how SMEs can benefit from Business Analytics, the challenges, and solution, this can be found in Coleman et al. (2008).[6]

The major solution to some of the problems stated earlier is to implement a data governance policy for the business no matter how small. Data governance is a set of procedures, responsibilities, policies, standards, and measurements that ensure that information is used effectively and efficiently to help an organization achieve its objectives.[7] Small businesses need to realize that the earlier they have this in place, the better they are prepared for the future that is here already.

1.6 Types of Analytics Problems in SME

In this section, we try to highlight some types of questions SMEs can answer with data analytics and some suggested techniques that might help to explore such (Table 1-2). It is important to note that these questions are informed by the kind of business and the nature of the analytics goal at hand. Also sometimes, there might be a need to combine the techniques or use them as prerequisite to each other as demonstrated in the consulting business problem scenario in Chapters 9 and 10. It is also important to note that the answer to these questions can only yield the needed value when embedded in the analytics process described in Section 1.3.

Table 1-2. *SME Problems and Techniques*

Questions	Suggested Techniques
• How does promotion embark my sales?	Time series forecasting
• Who buys our products during promotions or who are our customers?	Association rule mining, network analysis, collaborative filtering
• Which type of displays work better?	
• What products are selling together and what products are neutralizing each other?	
• How can we target our marketing and promotions and offer recommended products to customers?	
• How can we discover the most influential customers/prospect?	
• How can we know our employee efficiency?	
• What do we stock and how do we position them?	

(continued)

Table 1-2. (*continued*)

Questions	Suggested Techniques
• What are customers buying? • What are the categories of customers we have?	Clustering
• Why are our sales dropping or declining?	Descriptive/exploratory analysis
• How can we predict sales? • How can we predict the impact of discounts on sales? • How can we estimate the profit of a sales adventure, predict prices, etc.? • Which customer is going to buy or buy next? • How many purchases they are going to make? • How many of our customers will stay with us or leave? • How do we retain customers and target them in the right way? • How do we turn nonprofiting customers to more profiting ones or profiting to more profiting? • What do we store or discard? • When do we store or discard? • What is the correct price point that is maximizing sales? • What are the most profitable products? • How do we increase our customer base?	Classification and prediction techniques

1.7 Analytics Tools for SMES

Some of the analytics tools that small businesses can use are captured in Figure 1-3, though there are still more. Some of these tools are available as open source and can be used for free, while some are not. As tempting as it may seem for businesses to want to opt for the free ones, it is important to find out what is obtainable on these platforms before settling for them. In this book, we primarily use RapidMiner Studio but also augment it with RStudio and Gephi in order to have a robust view to the analytics procedure. RapidMiner Studio was selected so as to appeal to nonprogramming Business Analytics audience, and Gephi is used for the network analysis to create beautiful and easy-to-interpret networks. It is also important to note that all the tools used in this book are open source (available for free download online).

Figure 1-3. *Some analytics tools for SMEs*

1.8 Road Map to This Book

This book is divided into four major parts. Figure 1-4 gives the overall structure of the book and where the topics in this book fit in. Chapter numbers are indicated beside the topic. Part 1 (Chapters 1 and 2) explains the fundamental concepts explored in this book, such as data science, data science for business (Business Analytics), and what it takes to carry out any analytics project both generally and specifically for a small business. In this part, we also explore issues around data and how to manage and prepare it for the analytics project with practical examples. Part 2 (Chapters 3 and 4) focuses on analytics consulting and explains how to navigate your way through to becoming successful in the data analytics consulting space. It also gives a detail of the phases involved in Business Analytics consulting. Part 3 (Chapters 5–8) is focused on the data mining techniques common with small businesses, and this is expressed in an approach that first explains the basic concepts of these techniques in a simple way and then uses a real business problem scenario for the practical application. This part is practical oriented and based on case study problems experienced by small businesses. In this book, we will cover five different practical business problems. These business problems are covered in Chapters 5–8. The techniques used in the book demonstrate how to solve these problems. It is important to say here that despite using a particular problem as a case study, it is not only in this situation that the approach can be deployed, but it can be used in other similar problem scenarios. The techniques selected are based on their popularity in practice, and they fall under the broad category of prediction (predicting numerical outcome), classification (predicting categorical outcome), and descriptive analytics. Finally, Part 4 brings the consulting principles into practice by using an SME case study to model the already explained consulting phases in Part 2 and adopting the appropriate techniques among the ones explained in Part 3. Although each chapter stands alone, we advise that you read Part 1 before proceeding to Part 3, and Part 2 before proceeding to Part 4.

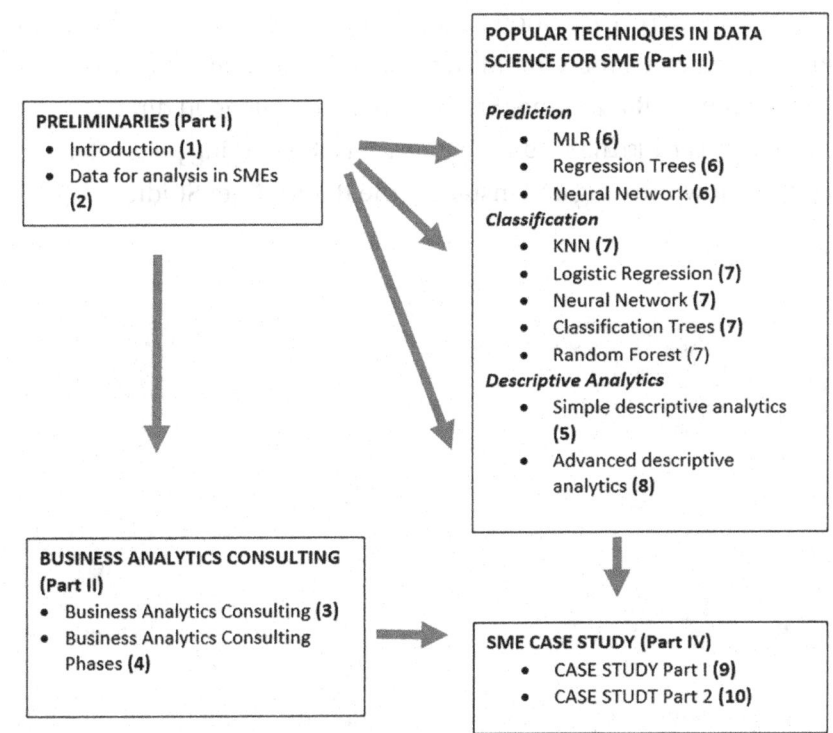

Figure 1-4. *Road map to the book*

Since this book is practical based, the following is a guide to installation and a brief introduction to the softwares we will be using for the practical demonstration.

Using RapidMiner Studio

RapidMiner Studio is a comprehensive open source data mining tool that contains over 400 built-in data mining operators as well as a broad spectrum of visualization tools. It was started by Ralf Klinkenberg, Ingo Mierswa, and Simon Fischer at the University of Dortmund, Germany. Today, it is currently maintained by commercial company plus open source developers. It comes in two editions, the community edition which is free and the enterprise edition which is commercial. The documentation and installation guide for RapidMiner can be found at https://docs.rapidminer.com/. RapidMiner Studio can be downloaded directly from https://rapidminer.com/. Figure 1-5 is the landing page after successfully downloading and installing the RapidMiner Studio.

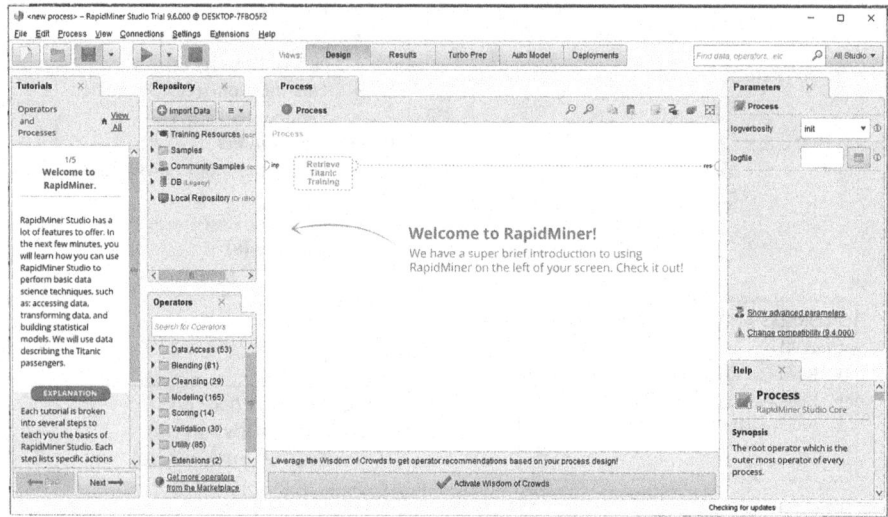

Figure 1-5. *Welcome to RapidMiner*

A detailed introduction to the RapidMiner Studio for first-time users can be found at https://docs.rapidminer.com/downloads/RapidMiner-v6-user-manual.pdf.

Using Gephi

Gephi is an open source and free software for visualization and exploration of all kinds of graphs and networks. To download Gephi, visit https://gephi.org/. Some installation guide can be found at https://gephi.org/users/install/. If you have any issues installing Gephi, visit www.youtube.com/watch?v=-JU-S5dMDVo. After installing Gephi successfully, you will see the screen in Figure 1-6. A quick introduction to Gephi can be found at https://gephi.org/tutorials/gephi-tutorial-quick_start.pdf.

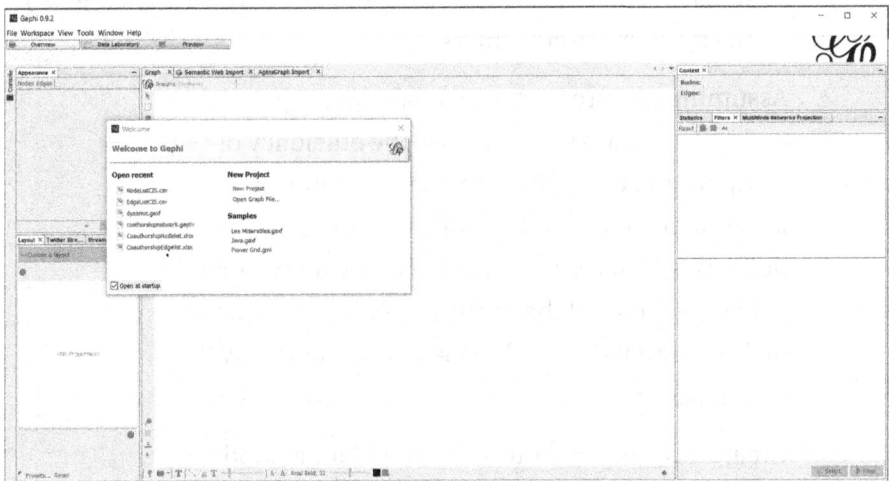

Figure 1-6. *Gephi screen*

1.9 Problems

1. Design a conceptual business model for the following business scenarios:

    a. A small business that intends to use information on its current training attendee to build a model to predict who will come for training in the future. The result of this prediction is to be used for target marketing.

    b. Hadetulah eye clinic is a small business of about 20 staffs in total and about five branches in Nigeria. They offer several services relating to eye treatment, from diagnosing to sales of eye treatment equipment. Even though they do not have

25

a core enterprise application running, they capture financial operations data in the form of excel sheets storing diagnosis results, information of the referral hospitals, equipment details, rental or sales information, and invoices. Their goal is to attract more customers.

2. Assuming you are part of an analytics project embarked on a task to improve the efficiency of an operation of an airline business. Your task is to understand how the business operates. In doing this, you are requested to come up with the conceptual model that gives a detailed explanation of the elements in that business space and how they relate together.

3. Create a conceptual business model for a noodle retail small business. This company intends to segment their current customers so as to be able to describe them for effective customer relationship management.

1.10 References

1. Galit Shmueli, Nitin R. Patel, & Peter C. Bruce, Data Mining for Business Intelligence, Concepts, Techniques and Applications in Microsoft Office Excel with XLMiner, published by John Wiley & Sons, Inc., Hoboken, New Jersey, 2010.

2. David Torgerson, Manuel Laguna, & Dan Zhang. Advanced Business Analytics Specialization [MOOC]. Coursera. www.coursera.org/ specializations/data-analytics-business

3. PwC's MSME Survey 2020 Building to Last, Nigeria report (June 2020), www.pwc.com/ng/en/publications/pwc-msme-survey-2020.html

4. Mara Calvello (January 28, 2020) What is Business Analytics and Why You Need It for Success, https://learn.g2.com/business-analytics

5. 10xDS Team (November 2019) Harnessing the power of Cognitive Analytics to reinvent your business, https://10xds.com/blog/cognitive-analytics-to-reinvent-business/

6. Shirley Coleman, Rainer Göb, Giuseppe Manco, Antonio Pievatolo, Xavier Tort-Martorell, and Marco Seabra Reis. How Can SMEs Benefit from Big Data? Challenges and a Path Forward, published online on Wiley Online Library.

7. Talend (March 2022) What is Data Governance and Why Do You Need It? (www.talend.com/resources/what-is-data-governance/)

8. www.gartner.com/en/information-technology/glossary/customer-analytics

9. www.techopedia.com/definition/29495/operational-analytics

10. Delmaster, R., and Hancock, M. (2001). Data Mining Explained. Boston: Digital Press.

11. Sylvain Giuliani (Feb 2022) What is Operational Analytics (and how is it changing how we work with data)? https://blog.getcensus.com/what-is-operational-analytics/

12. KPMG, Emerging Trends in Infrastructure. `www.kpmg.com/emergingtrends`

More resources on the chapter for further reading

- Eric Bradlow et al., Business Analytics Specialization [MOOC]. Coursera. `www.coursera.org/specializations/business-analytics`

- Installing RapidMiner, `www.youtube.com/watch?v=9YX66d192gY`

- Installing Gephi, `www.youtube.com/watch?v=-JU-S5dMDVo&t=2s`

CHAPTER 2

Data for Analysis in Small Business

In this chapter, we will look at the various sources of data generally and in small business. This is important because the major challenge of consulting for small business is the lack of data or quality data for analysis. This chapter will therefore detail the sources of data for analysis explaining first the type or form that data exists and some general ideas of how to collect such data. It gives an overview on data quality and integrity issues and touches on data literacy. In addition, we will explain typical data preparation procedures for data analytics techniques. To conclude, we look at data visualization, particularly toward preparing data for various analytics tasks as explained in Section 1.3.

2.1 Source of Data

Data can be retrieved in different forms; the most basic category of data is the structured, semistructured, and unstructured data.[1]

Structured data: This type of data has a predefined order to it, and it is formatted to a particular structure before it is stored. This structure is referred to as schema-on-write. One good example is data stored in a relational database management system.

© Afolabi Ibukun Tolulope 2022
A. I. Tolulope, *Data Science and Analytics for SMEs*,
https://doi.org/10.1007/978-1-4842-8670-8_2

Unstructured data: This is when data is stored in its native format and not processed until it is used, which is known as schema-on-read. It has no form or order. It can be stored in a variety of file formats; examples are emails, social media posts, presentations, chats, IoT sensor data, and satellite imagery.

Semistructured data: This form of data includes metadata that outlines specific features of the data[3] and allows it to be manipulated more efficiently than unstructured data. Data stored in XML format is an example of this kind of data.

In business, the types of systems that capture data include the following:

- Core enterprise systems

- Customer and people systems

- Product and presence systems

- Technical operations systems

- External sources systems

It's important to note that sometimes these systems overlap in their functionalities stated as follows in some situations.

Core enterprise: These include systems that help in managing the financial operations of the business. They are used for billing and invoicing, enterprise resource planning, supply chain management, purchasing and selling activities, monitoring production activities, and many more. When compared to the other classes, they have a vast volume and are normally created, saved, and maintained within the operating application.

Customer and people systems: These include client services, marketing automation, campaign management, human resource systems, and customer relationship management (CRM) systems.

Product and presence systems: They help to store information about product management. These include product management, content management, web management, and analytics. Many of the actions that people undertake in ecommerce can be measured in the most minute detail. These include customer reviews, weblog data, mouse click trails, global positioning satellite (GPS), and so on.

Technical operations systems: They are very tactical and help to monitor processes of other systems to make sure that they are functioning properly. They include process monitoring, alarming and fault monitoring, ticketing and workflow management, telematics, and machine data processing.

External source systems: These include data collected outside the organization but are relevant for the analytics process. They include demographic and segmentation data, data from partners, supplies and government agencies.

To extract this data for analytics or prepare it for secondary analytics storage like the data warehouse, there are several languages available which include but are not limited to SQL, Python, and so on. Sometimes, when the data is not available for extraction, we might need to collect it using either the active data collection approach or the passive data collection approach.

Active data collection: This occurs when data is collected directly from the user, for example, a user deliberately shares personal data when completing an online retail transaction. When collecting such data, oftentimes, there is a need to include some disclosure of the data being collected. Active data can also be collected using surveys whether online or offline, and there are procedures to be followed[2] to carry this out to achieve a particular goal. Online platforms include Google Forms, SurveyMonkey, and so on.

Passive data collection: This occurs when the object of interest is not involved in the data collection. It is collected through scanner or media such as radio, television, and social media. Other passive means of collecting data include web and mobile platforms. Web data gives you a lot of information about what your customers are looking at either on your own website or even the competitors' website. For ecommerce websites, for example, we can collect customer reviews, web transaction data, customers' demographic data, and weblog data. An example of companies that help collect web data is Comscore (`www.comscore.com/`). These companies can help to collect data which can give insights to how a particular company is trending on social media, see what is happening on the company's competitor's website, and so on.

A real business illustration I: Consulting for small business requires creativity in sourcing for analytics data; for example, a small business that organizes training on different subjects aims at using information on its current training attendee to build a model to predict which of its prospects will be interested in training in the future. Their intention is to approach the prospects predicted to be interested with their new training package. Since it's a small business, there is minimal budget to carry out this task, and therefore they need to know the exact prospects to target with the available resources. In order to get external data for this task, the data analyst had to put a targeted advert on Facebook, accompanied by a Google Form link which is used to collect the details of interested prospects. This is one way to source external data creatively.

A real business illustration II: Hadetulah eye clinic is a small business of about 20 staff in total and about five branches in Nigeria. They offer several services relating to eye treatment; from diagnosing to sales of eye treatment equipment. Even though they do not have a core enterprise application running, they capture financial operation's data in the form of excel sheets. They also store information on the referral hospitals, equipment details, rental or sales information, and invoices. For Hadetulah, the customer and people systems are captured in excel sheets in the form of name of customer, phone, age, sex, occupation, address, and so on. They are also able to capture information such as the amount spent on types of lenses, staff who attended to the customer, and so on. Figure 2-1 gives an illustration of a feature representation example for Hadetulah opticals.

From this figure, we can see the customer's picture and the features or attributes that represent the customer. The figure further gives an example of the values that each attribute might have and illustrates how this (an historical collection of features of different customers) will be an input to the prediction algorithm given that the attribute amount is the label. After developing the model, we can now use it to predict for a new customer.

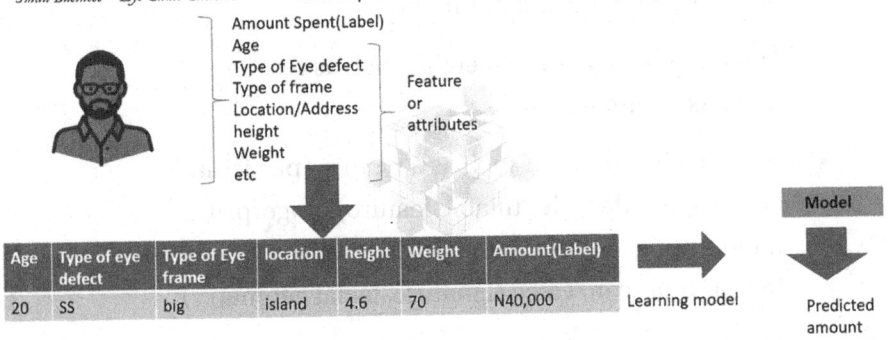

Figure 2-1. *Feature representation of Hadetulah optical*

Data Privacy

It is very important as a data analyst to familiarize yourself with the data privacy guidelines, and ignorance is not an excuse even when it comes to data privacy issues. To this end, there are standards in place to help enforce data privacy. Professional or commercial organizations establish certain ethical standards in order to put in place a set of nonlegally binding information treatment procedures. To enforce the treatment of certain sets of data, legal standards have been established by law and order. A company or an institution can also establish its own policy guidelines or standards. There is also the good judgment standard, which states that even if something is technically correct by more formal rules, one should constantly consider "Is this a good idea?" and "What might the repercussions be?"

The following are the privacy guidelines that need to be followed:

- Avoid using complete names, maiden names, mother's maiden names, or nicknames.

- Do not include personal identification numbers, National Identity Numbers, passport numbers, driver's license, taxpayer identification numbers, or financial account or credit card numbers.

- Do not include address details, such as residential address or email address.

- Personal information such as a picture (particularly of a face or other identifiable feature), fingerprints, handwriting, or other biometric data should be avoided (e.g., retina scan, voice signature, facial geometry).

- Do not include information about a person that is linked or linkable to one of the preceding information (e.g., birth certificate, birthplace, race, religion, weight, activities, geographical indicators, employment information, health records, academic information, financial details).

There is a need to find out the data policy operable in your own country; in Nigeria, for example, we have Nigeria Regulations on Data Privacy.[4]

2.2 Data Quality and Integrity

It is often said in the world of computing that garbage in is garbage out, that is, the quality of information produced after processing cannot be better than the quality of information that was processed. This is not any less true in data analytics. The quality of data refers to the degree to which data can be used for its intended purpose and the degree to which data accurately represents the real world. There is a need to measure the quality of data to be used for analysis, and this can be done by checking the data for the following:

Completeness: This is used to check if we have all of the information we should have. We respond to questions such as, "Have all events been captured?" Are all reference data values taken into account?

Uniqueness: This is when we check if single events are captured only once.

Accuracy: This is where we determine if the data is a true representation of the concept it is attempting to convey. Ensuring numerical and string values are correct, as well as timestamps and other properties, is among the other activities.

Consistency: In checking for consistency, we want to ensure that the data format is as expected. There is a need to ensure that the same data is captured the same way every time.

Conformance/validity: Does the data in the database comply with the syntax, encoding, or other model requirements? Are data formats correct? Are codes as expected? Are naming conventions adhered to?

Timeliness: Here, we want to know if data is available by the time it is needed. This is also referred to as "data latency."

Provenance: Here, we want to know if we have visibility into the origins of the data. And how much confidence do we have that the data is real and accurate?

From a survey in 2017 on 7376 respondents by Kaggle[5] on the data science competition community, it was discovered after asking them to rate the barriers they faced at work that the highest in the rank was the dirty data problem which carried a percentage of 49.4% out of the whole population surveyed.

Particularly common in the SME business environment is the *dirty data problem*. These are situations where data is full of errors such as misspelled data, typos, and duplicate data. This can be fixed systematically when identified. There are also situations where data violates business rules; to solve this, there might be a need to involve a cross-functional effort between departments of the organization. There are situations where data is collected using the wrong method or the wrong population. This data often comes from asking the wrong business questions; all these can be overcome by putting a proper data governance procedure in place for the business at the start or carry out corrective measure when discovered. Currently, there are algorithms that can be used to analyze data for duplicates and other mistakes. It is important to emphasize here that prevention is the best cure, that is, improved procedures of data collection, forms, and records with more standardized fields are preferable. In addition, make origins and history of data visible and transparent in order to trace back every mistake. In conclusion, the completeness, validity, and consistency of your data depend on the methods, and each business will have developed and customized such methods for its particular needs.

2.3 Data Governance

Data governance is simply defined as the process put in place to make sure that data is available, usable, secured, and has integrity. In the past, only large corporations cared about data governance, but because small businesses are also onboard making data-informed decisions, they are also embracing data governance. The data governance process is built on a collection of rules and regulations that secure data and ensure that it is handled in accordance with all applicable external laws. The data steward is responsible for developing and enforcing these uniform norms and regulations. For small businesses, the way to go about data governance is to tackle it from these three directions. First, assign someone (or more if you can afford it) to the task of data governance. A good rule of thumb is that as soon as you have a full-time analyst in your team, they should explicitly own data governance.[9] Second, the person in charge comes up with a document known as a data *treasure map* consisting of a list of where data comes from, for example, marketing, finance, sales CRM, etc., and a list of where the data goes to, for example, data warehouse if any or the business intelligence visualization tools. (Note that small business also refers to one-man business, so this second stage might not be available in the form of a software application but interpreted as copying data from Excel to Tableau manually). Third thing is to include the information about who uses your data. Finally, there is a need to educate all employees on the importance of the data governance model developed, its dos and don'ts. For more resources on data governance for small business, check Nwabude, Begg, and McRobbie (2014).[6,7] For the tools that can help small business with data governance check Nwabude(2014) and others[8,10]

2.4 Data Preparation

After the necessary data has been retrieved for analytics, the very first thing to do is to clean the data. This is necessary because analysis done on faulty data can be misleading, and there can be serious economic consequences of using uncleaned data for analysis. In the past, many major organizations have lost a lot of revenue due to this, and a popular example is the JPMorgan Chase trading loss.[11]

First of all, let's lay some foundation about the types of variables so that we can understand the data preparation procedures. The data obtained is in the form of variables. Categorical variables have a fixed range of values and can be unordered (called nominal variables) with categories such as North America, Europe, and Asia; or they could be ordered (called ordinal variables) with categories such as high value and low value. Numerical variables are never ending. They are continuous (able to assume any real numerical value, usually in a given range). Numerical variables can be integers which take only integer values. Another kind of variable is the date, and we have other ones depending on the analytics platform being used.

When cleaning the data, we are examining the data for bad formatting, trailing spaces, duplicates, empty rows, synonyms, and abbreviations. In cleaning the data, you ask questions like: *Do character variables have valid values? Are numerical values within range? Are there duplicate values/ rows? Are values unique for some variables? Are dates valid? Does the dataset even contain attributes indicating what you want to predict? In which case there is no need for prediction. Are there missing values? Is there inconsistency in description?*

In preparing and examining the data you want to use, summary statistics can be used to discover the complete set of variables in the data, learn the characteristics of the variables and the range of values in the variables, and select variables for the analysis. Mean, median, standard deviation, minimum, maximum, and so on are all examples of summary statistics. The minimum and maximum can be used to detect extreme values that might be errors. The mean and median give a sense of the central values of that variable, and a large deviation between the two also indicates skew. The standard deviation gives a sense of how dispersed the data are (relative to the mean).

Summary Statistics

Example 2.1

The data *OnlineQuestion1.xlsx* needed to go through a data cleaning process. First, understand and familiarize yourself with the data using summary statistics, handle missing data and outliers, and finally normalize the data. We will deal with each of these tasks as they are treated.

1. The first thing to do is to create a repository for your project using the following steps:

    • Create a repository; click the arrow down beside **Import Data** in Figure 2-2.

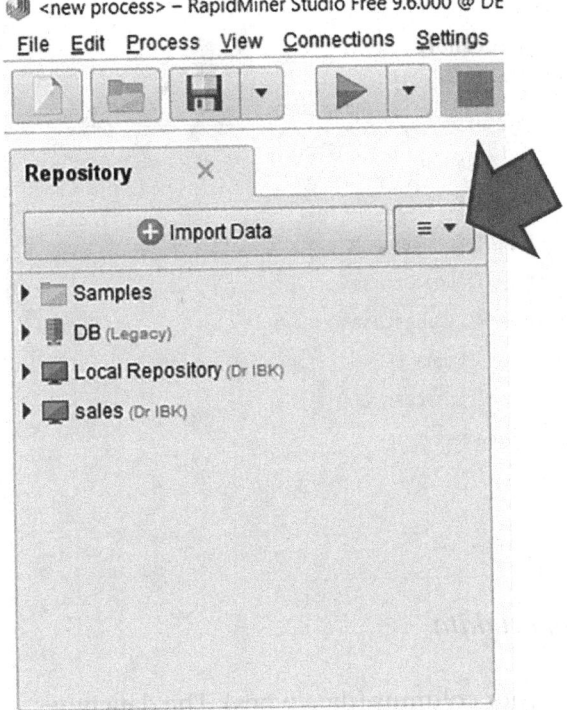

Figure 2-2. *Create a repository*

    • Give your local repository a name "*OnlineCourse.*"

        • Create two folders in your repository (data and process).

2. Load the data to be cleaned.

- Click **Import Data** (Figure 2-3).

- Select your data file (*OnlineQuestion1.xlsx*) from anywhere it is on your computer.

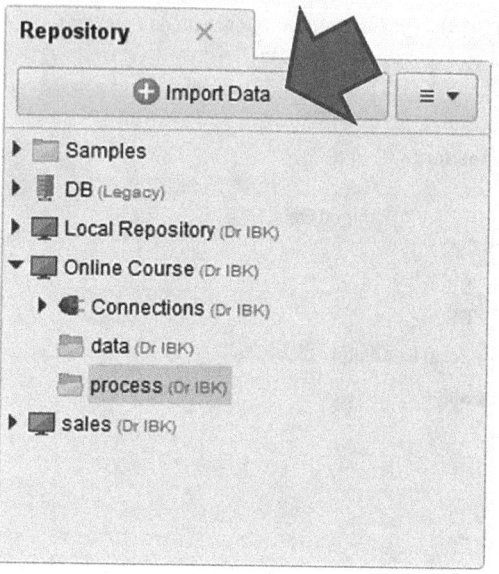

Figure 2-3. *Import data*

3. Format your columns (data types). The data types in RapidMiner are polynominal (categorical variables with more than two categories), binominal (categorical variable with two categories), real (numbers with decimals), integers (numbers without decimals), date_time (data and time together), date, and time.[13] At this stage (Figure 2-4), you can examine all attributes to see if their data type is in order based on domain knowledge and the purpose of the analysis. This is when you answer

questions like: *Are values unique for some variable?*
etc. There might be a need to change the type,
change the role of the variable, rename the variable,
or remove the variable (this can be done by clicking
the arrow beside the attribute in Figure 2-4). For
this purpose of demonstration, let's change the
attributes *message* and *MOVED_AD* to binominal.

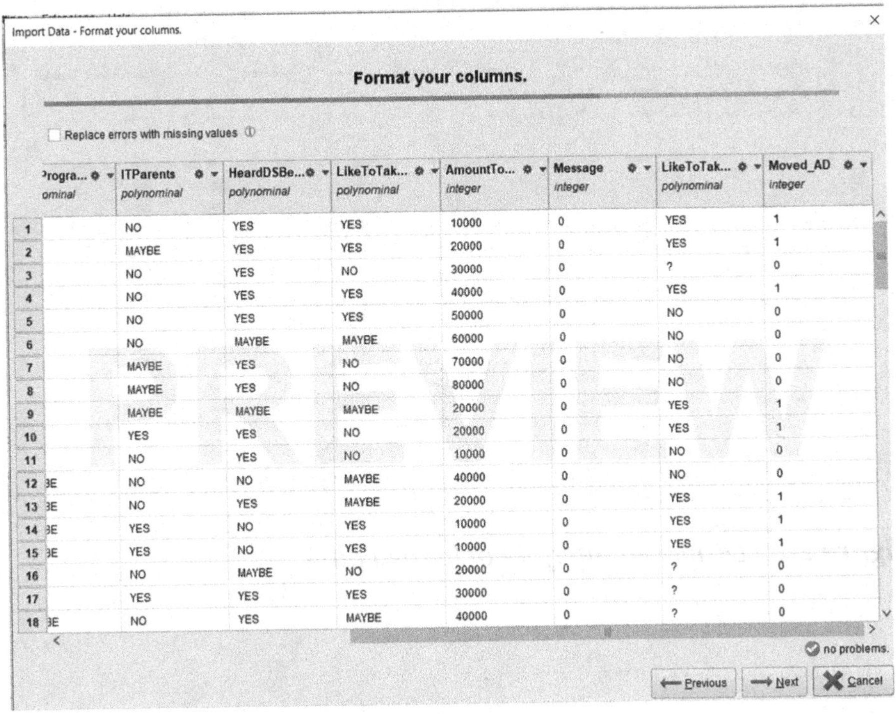

Figure 2-4. *Format data types*

4. In situations where there might be errors (e.g.,
    character variable having numerical data and vice
    versa), the *Replace error with missing values* will
    take care of this. In the next step of the cleaning
    process, we can deal with the missing data.

5. To load/import the data into the repository, select the data subfolder in your repository and click ***Finish***.

6. The displayed interface of loaded data is revealed in Figure 2-5. From here, you can click the ***Statistics*** to reveal the summary statistics (Figure 2-6).

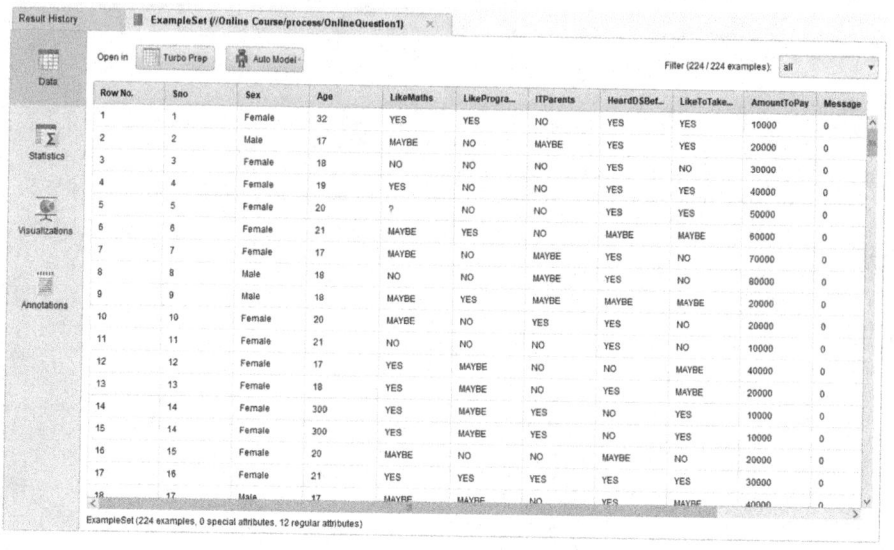

Figure 2-5. *Displayed interface of loaded data*

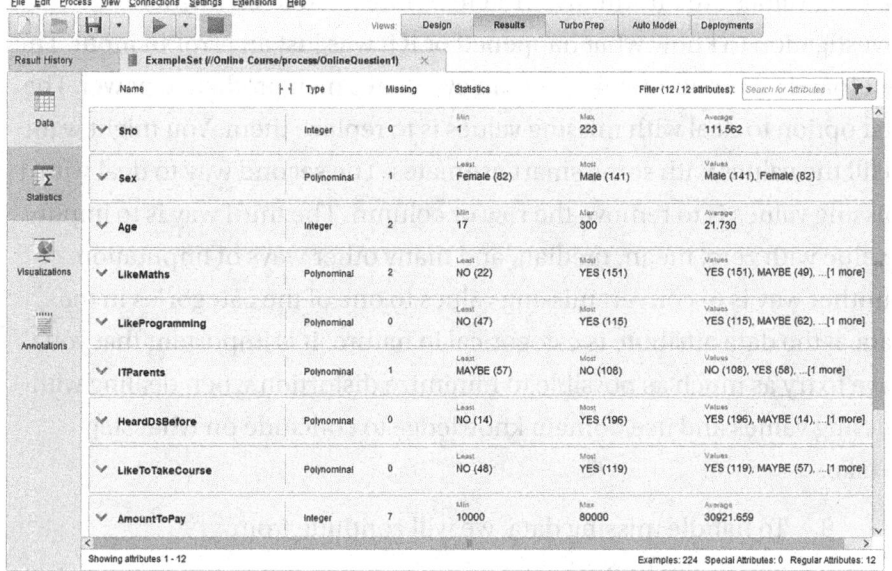

Figure 2-6. *Summary statistics*

7. Examine the statistics to solve the following issues:
    Does character variable have valid values? Are
    numerical values within range? The min and max
    functions can be used to detect extreme values that
    might be errors. Are there missing values? Check
    for the following: difference in scales, definitions
    of fields, units of measurement, time periods,
    and so on.

Missing Data

How do we handle missing data? You cannot just simply throw away data
in many cases because you need to have enough data for meaningful
analysis. Also, there might be more than one attribute with different
missing values in which case too many rows will be removed. Sometimes,

missing values are informative. You therefore need to do further investigation to know what happened or if it was just an error in input. The fact that data points are missing can even have high predictive power. The first option to deal with missing values is to replace them. You might want to fill the values with some smart estimates. The second way to deal with missing values is to remove the row or column. The third way is to impute a value with zero, mean, median, and many other ways of imputation. Another way is to convert missing values to one of the categories in the data, if the data attribute is categorical in nature. It is important that you have to try as much as possible to minimize distortion when dealing with missing values and use domain knowledge to conclude on what step to take.

8. To handle missing data, we will continue from where we stopped in step 7.

- From Figure 2-6, click the ***Design*** tab and drag the data to the design view from the repository (this is how we add the operator or data to the design view).

- Connect the output of the data to the result (**res**) as seen in Figure 2-7.

- Run the processes by clicking the ***run*** (blue triangle on the tools bar) button (indicated in Figure 2-7).

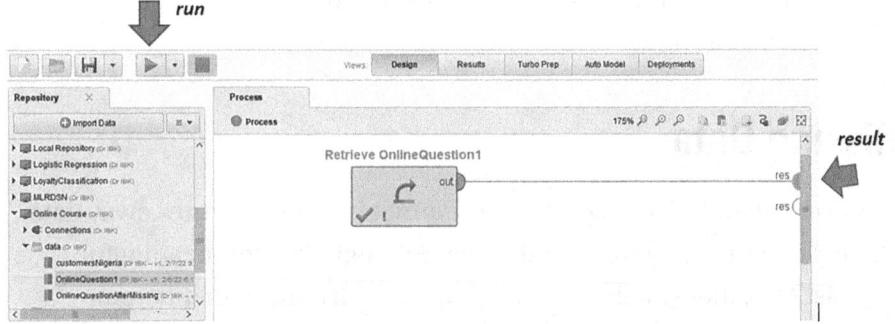

Figure 2-7. *Running your first process in RapidMiner*

- From the output after clicking **Statistics** (Figure 2-8a),
  you will observe that not all attributes have missing
  values, and this is where you decide how to deal with
  the missing values.

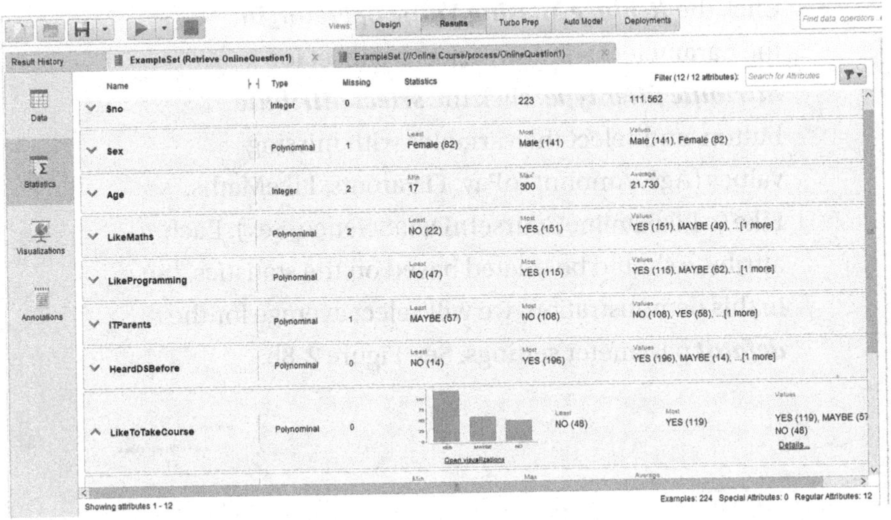

Figure 2-8a. *Checking the statistics again*

- We cannot use the **Use Filter Examples** operator to
  remove missing values because that would amount to
  over 50 records being lost, so we will use the **Replace
  Missing Value** operator.

- Under **Operators**, search for the **Replace Missing
  Value** operator and add it to the design in Figure 2-7
  and link it to the **Retrieve OnlineQuestion1** operator
  and then to the output.

- At this point, it is important to know that there are
  different ways to handle different attributes in the
  parameter settings of the **Replace Missing Value**
  operator. If you want to deal with the attribute one by

45

one, you can first use the *select attribute* operator to select the attribute you want to deal with and then use the *Replace Missing Value* operator to deal with the attribute.

- Click the *Replace Missing Value* operator; in the parameters window, select a subset for *attribute filter type*, click the *select attribute* button, and select the variables with missing values (Age, AmountToPay, ITParents, LikeMaths, LikeToTakeOnlineCourseInDataScience, Sex). Each attribute should be treated based on the statistics, but in this demonstration, we will select average for the *default* parameter settings. See Figure 2-8b.

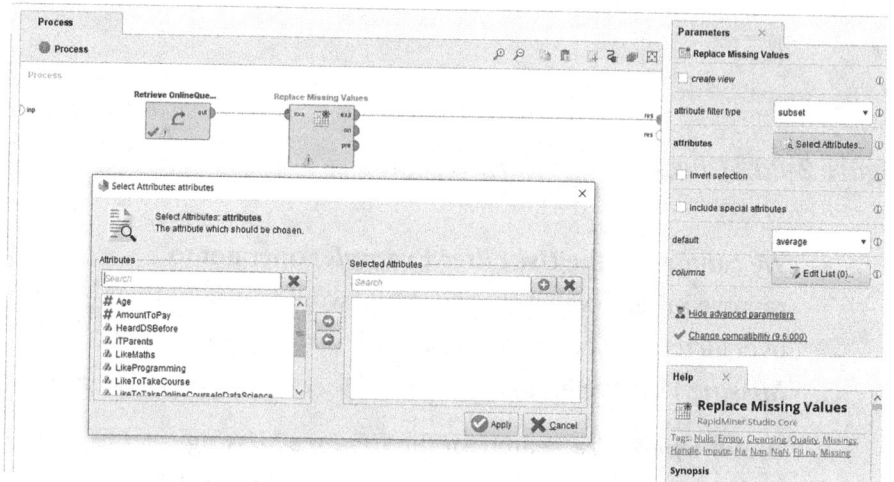

Figure 2-8b. *Replace Missing Values operator*

- After running the process, all the missing values should be corrected when you check the statistics. You can also click the *Results* button to check the output after dealing with missing values.

- Other data cleaning activity is to check for duplicate roles as revealed in lines 14 and 15 of your output. To solve the problem, add the **Remove Duplicates** operator to your design. To save the data to a CSV file (which we will use for the next stage of data preparation), add the **Write CSV** operator to your design as shown in Figure 2-9 (create a new csv file and name it *OnlineQuestionAfterMissing* and put it on the desktop), click the **Write CSV** operator and select the csv file you have created earlier, and run the process. If you open the csv file, you will see the data you have cleaned so far. You can also note in the output that the duplicate row has been removed.

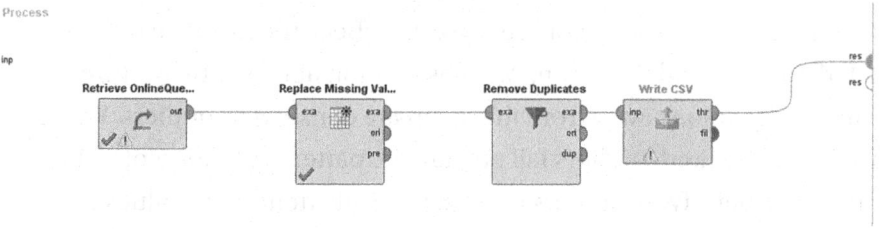

Figure 2-9. *Writing to a CSV file*

Data Cleaning – Outliers

What are outliers? An outlier is an observation that lies an abnormal distance from other values in a random sample from a population. To detect outliers, we check if values are in a reasonable range, given what you would expect for each variable. When there are obvious outliers, for example, in Figure 2-10, there is a need to deal with them.

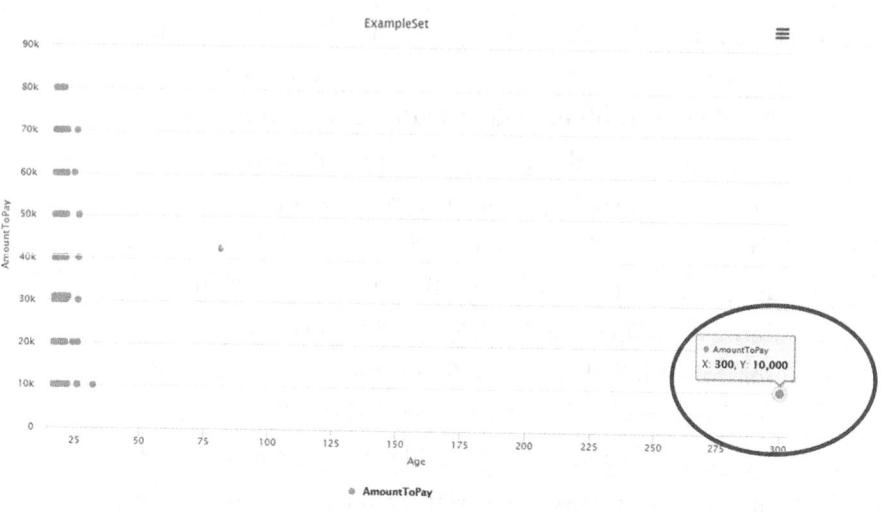

Figure 2-10. *Outliers*

Furthermore, to detect outliers, we can check the distance between each data point and the mean. Any observation above or below three standard deviations away from the mean are considered outliers. When variables are paired, outliers fall outside the pattern (scatter plot). The purpose of identifying outliers is usually to call attention to values that need further review, for example, a temperature of 200°F for a sick person or age of 300. Some key questions about outliers are: Is the outlier a mistake or legitimate point? Is the outlier part of the population of interest? The answer will help to decide whether to include the outlier in our analysis or not. One way to detect outliers is to sort the records by the column, then review the data for very large or very small values in that column. You also look at each column's minimum and maximum values (this can be done using statistics in RapidMiner). Boxplot visualization or a scatter plot of two numerical variables can also be used to find outliers. If the outlier is due to a mistake, fix it; if the outlier is outside the target population, eliminate it. If outliers are much beyond the range of the rest of the data (e.g., a misplaced decimal), it may have a significant impact on some of the data mining processes we plan to employ, such as Euclidean distance in clustering. Other methods for dealing with outliers include data transformation, the use of outlier-resistant tools such as mean instead of median, and so on. All of these are decisions that should be made by someone with domain expertise of the application in question.

9. We continue our practical example by dealing with outliers. Create a new process, and *Import* the data (*OnlineQuestionAfterMissing.csv*) resulting from the missing value cleaning, using the procedure explained in step 2 of this example. To take a look at the data before handling outliers, you can run the process with only the data in the design view. To handle outliers by filtering out the row, use the process in Figure 2-11; click the *Select Attributes* operator; for *attribute filter type*, click *subset* and select all the attributes so the output of the process will contain the whole data stored. Note that, sometimes, you might want to normalize before applying the *detect outliers* operator, particularly when using Euclidean distance. For the *Detect Outliers* operator, note that in the parameter, the number of neighbors is set to 10 by default. You might want to change that based on your understanding of how the operator works. Just as it is with all other operators in RapidMiner, there is a need to read the help document of all the operators to know how to maximize it for your own particular task. The *Filter Examples* operator selects which row(s) of the data is kept and which is removed based on how you program it. You can set the *filters parameter*, thus outlier> equals >false; if you do this, you will observe that even though we expected

only one row to be removed, more roles are gone.
A more efficient way to deal with this problem
is to handle outliers by correcting rather than
deleting the row. Finally, the **Write CSV** operator is
used to save the resulting data in a csv file named
OnlineQuestionAfterOutlier.csv.

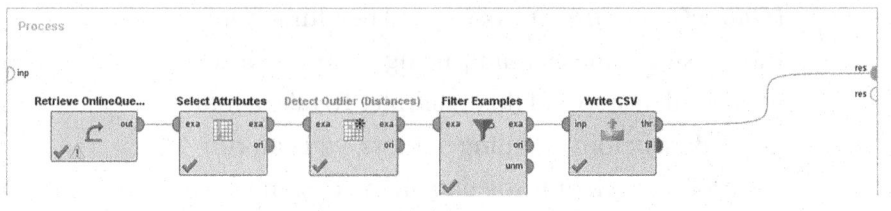

Figure 2-11. *Process for filtering outliers*

10. Instead of removing the entire row as in step 9, you
can filter the results based on specific attributes.
To do this, you first need to detect the number of
outliers through visualization. For example, in
this case we have been able to detect that we have
one outlier, and we see that the value is >=300 for
attribute age. To deal with the outlier based on
this information, you can use the same process in
Figure 2-11 and then for the filter examples operator
set as in Figure 2-12a.

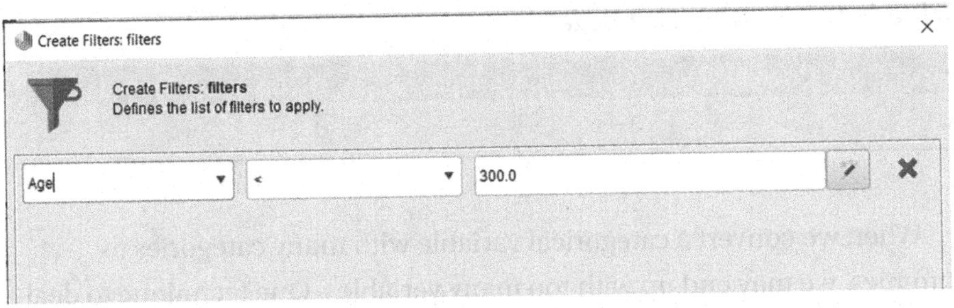

Figure 2-12a. *Correcting outliers*

Normalization and Categorical Variables

Normalization is used to scale attribute data to fall within a specified range. It is the method of organizing data to appear similar across all records and fields. We normalize so that attributes will not take undue advantage over each other in modeling or analytics processes. Some algorithms require it, for example, neural networks; also, some clustering algorithms require data normalization. There are several types: min-max normalization, z-score normalization, and so on. Min-max normalization is one of the most common ways to normalize data. For every feature, the minimum value of that feature gets transformed into a 0, the maximum value gets transformed into a 1, and every other value gets transformed into a decimal between 0 and 1.

Handling Categorical Variables

When dealing with ordered categorical variables (age group, degree of creditworthiness, etc.), label the categories numerically (1, 2, 3, ...) and handle the variable like it were a continuous variable. If a categorical variable is nominal, sometimes there is a need to convert it to dummy. For example, if occupation is a categorical attribute with "unemployed," "employed," or "retired" categories and a record has a category of unemployed, this is translated as in Table 2-1.

Table 2-1. *Creating Dummies*

Occupation_Unemployed	Occupation_Employed	Occupation _Retired
1	0	0

When we convert a categorical variable with many categories to dummies, we may end up with too many variables. One technique to deal with this is to combine close or similar categories to reduce the number of categories. Combining categories involves the use of both professional knowledge and common sense. In general, categories with a small number of observations are suitable candidates for combining with others. Only use the categories that are most important to the analysis and mark the others as "other." You can also use a bar chart to make such decision.[12]

1. To normalize the data resulting from the outlier treatment OnlineQuestionAfterOutlier.csv, you can use the ***Normalize*** operator, and to convert to a dummy, you can use the ***Nominal to Numerical*** operator (see Figure 2-12b). For both the ***Normalize*** and ***Nominal to Numerical*** operators, you will need to select the attributes using the ***attribute filter type*** and ***attribute*** parameters as shown in Figure 2-12b.

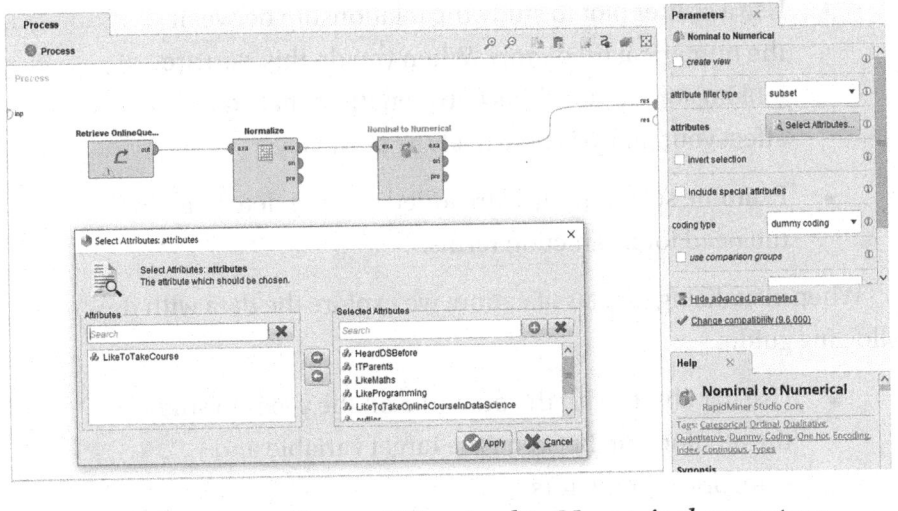

Figure 2-12b. *Normalize and Nominal to Numerical operators*

2.5 Data Visualization

Data can be visualized for several purposes in an analytics project.
Data can be visualized for prediction, classification, and unsupervised
learning, also known as advanced descriptive analytics. For a detailed
understanding of data visualization, check out the following references:
Galit et al. (2019), chapter 3 or Thomas Rahlf (2017).

When visualizing data for prediction, we explore the data with the
following guide:

- Use a histogram/boxplot to determine any needed
  transformation (plot the outcome variable against the
  numerical predictors). If it's the histogram, for example,
  we check skewness. This will be covered in Chapter 6.

- We can also study the relation of the target variable
  to categorical predictors via boxplots, bar charts, and
  multiple panels.

- Use a scatter plot to study the relationship between the numerical predictors. When you do this, you are looking for relationships between them that could affect your analysis negatively or positively.

- Examine scatter plots with added color to determine the need for interaction terms.

When visualizing for classification, we explore the data with the following guide:

- Use bar charts with the outcome on the y axis to study the relationship between the target variable and categorical predictors.

- Use color-coded scatter plots to study the relationship between the target variable and two numerical predictors to denote the outcome.

- A side-by-side boxplot, in which each numerical variable is compared to the outcome, can be used to investigate the relationship between outcome and numerical predictors. It's worth noting that the most separable boxes represent predictions that could be useful.

Use the following as a suggestion when visualizing for an unsupervised task:

- Create scatter plot matrices to identify pairwise relationships and clustering of observations.

- Use heat maps to examine the correlation table.

- Use various aggregation levels and zooming to determine areas of the data with different behavior.

2.6 Problems

1. Using the data named OnlineQuestionAfterOutlier. csv, create dummies of all categorical attributes and normalize all numerical attributes.

2. Perform the typical data preprocessing for prediction for the data named FarmCo.External. xlsx; describe the steps you took for this process, conclusions made, and the resulting data (the target attribute is Unit Price).

2.7 References

1. Talend, (2022, March) Structured vs. Unstructured Data: A Complete Guide, `www.talend.com/resources/structured-vs-unstructured-data/`

2. Online Survey Design and Data Analytics: Emerging Research and Opportunities (Advances in Data Mining and Database Management) by Shalin Hai-Jew, published by IGI Global, 2019.

3. Feature Engineering for Machine Learning: Principles and Techniques for Data Scientists by Alice Zheng, Amanda Casari, published by O'Reilly Media, 2018.

4. S.P.A. Ajibade & Co, (2020 Febuary) Data Privacy and Protection under the Nigerian Law – Francis Ololuo, `www.spaajibade.com/resources/data-privacy-and-protection-under-the-nigerian-law-francis-ololuo/`

5. www.kaggle.com/kaggle/kaggle-survey-2017

6. Evan Kaeding and Pinja Virtanen (July 2021) Data governance for startups and SMBs: what it is and why you should care, https://supermetrics.com/blog/data-governance

7. Nwabude, C. B. (2014). Data Governance in Small Businesses – Why Small Business. *2014 3rd International Conference on Business, Management and Governance* (pp. 101–107). IACSIT Press.

8. Best Data Governance Software for Small Businesses, www.capterra.com/data-governance-software/s/small-businesses/

9. Data Governance: An Essential Guide – 5 principles, 10 components, and getting started, https://satoricyber.com/data-governance/essential-guide/

10. Best Data Governance Software for Small Businesses, www.g2.com/categories/data-governance/small-business

11. 2012 JPMorgan Chase trading loss, https://en.wikipedia.org/wiki/2012_JPMorgan_Chase_trading_loss

12. Galit Shmueli, Nitin R. Patel, & Peter C. Bruce, Data Mining for Business Intelligence, Concepts, Techniques and Applications in Microsoft Office Excel with XLMiner, published by John Wiley & Sons, Inc., Hoboken, New Jersey, 2010.

13. https://docs.rapidminer.com/

More resources on the chapter for further reading

- RapidMiner: Data Mining Use Cases and Business Analytics Applications by Markus Hofmann & Ralf Klinkenberg, published by CRC Press, 2014.

- Excel Data Analysis: Your visual blueprint for analyzing data, charts, and PivotTables, 4th Edition, by Paul McFedries, Released July 2013, published by Visual.

- Graphical Data Analysis with R by Antony Unwin, published by CRC Press, 2015.

- Data Analytics for Business Specialization Course on Coursera.

CHAPTER 3

Business Analytics Consulting

In this chapter, we will look at Business Analytics consulting, particularly what the concept implies and how to build such a career path. We will explain the types of Business Analytics consulting that exist and then narrow them down to how to navigate the world of Business Analytics consulting for small businesses. In this chapter, we will look at how to manage a typical analytics project and measure the success of analytics projects. In conclusion, we will discuss issues revolving around how to bill analytics projects, particularly as a consultant.

3.1 Business Analytics Consulting

A consultant is someone who is engaged in the business of giving expert advice to people working in a professional or technical field. Business Analytics consulting therefore refers to when such expert advice entails any stage of the Business Analytics journey as described in Chapter 1, Section 1.3. It can also be seen as the act or process of educating clients on varying aspects of data and modern technology involved in taking data, understanding it, processing it, and extracting value from the data.[2]

© Afolabi Ibukun Tolulope 2022
A. I. Tolulope, *Data Science and Analytics for SMEs*,
https://doi.org/10.1007/978-1-4842-8670-8_3

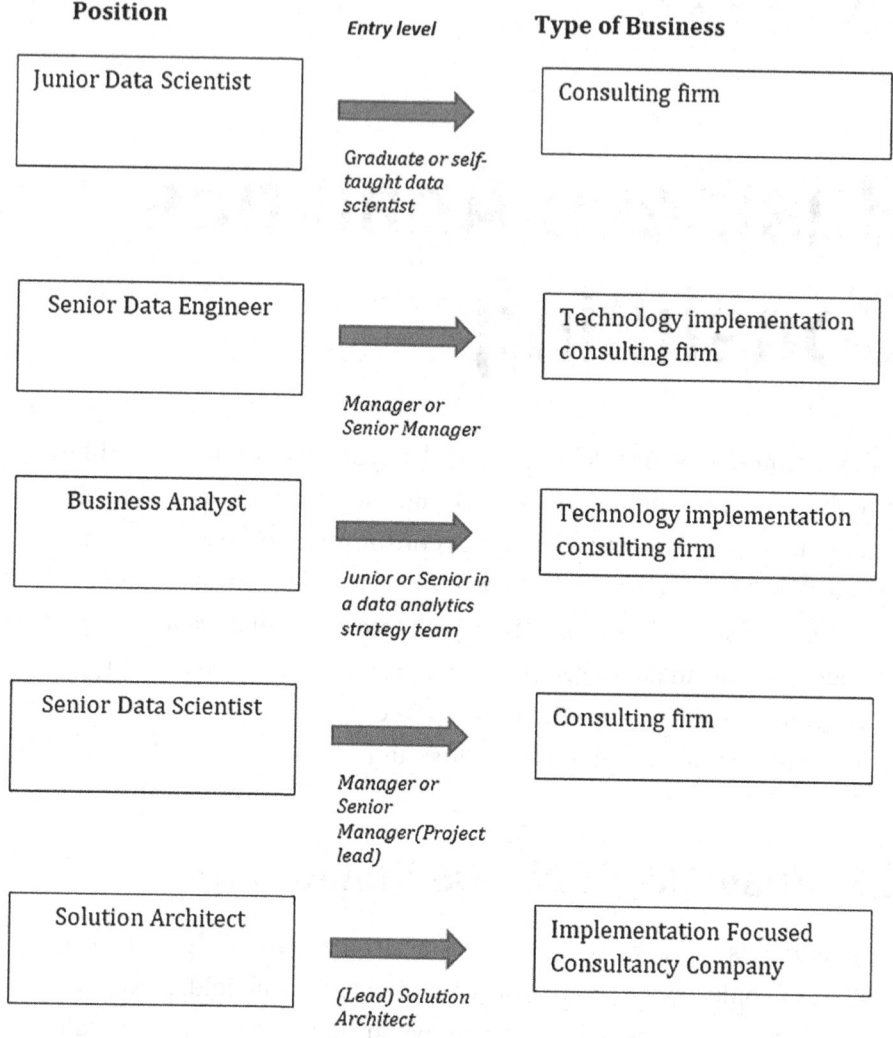

Figure 3-1. *Data scientist positions*

Figure 3-1 gives some examples of different positions and entry-level roles of data scientists in different types of organizations. From Figure 3-1, a graduate or a self-taught data scientist can secure a position as a junior data scientist in any consulting firm and so on.[1] Today, these roles might be customized depending on the type of large company in question.

Some of these companies also take data science interns. There are also independent data science consultants who offer data science consulting services without working for the company or organization whether it's a small or large business. Some are referred to as data science freelancers, and to get jobs, they use platforms like www.freelancer.com/, www.toptal.com/, www.upwork.com/, and so on, while some just position themselves as Business Analytics consultants or consulting firm and work independent of any organization or platform. In practice, there are several issues that an independent data scientist might face; on the top of this list is the resistance of the business in question to release all the data needed for analysis. The advice to independent data science consultants is to not scare the business by asking for all you need at once. However, you have to do your initial analysis and know what you will need to solve the problem. The approach is to retrieve some amount of data to start with and show some results that can encourage the clients to open up more in terms of data. You can simulate what they stand to gain from the data if made available. Always carry your clients and stakeholders along at every stage of the analytics project. Business Analytics consulting should be seen as offering products, in which case the product is in the form of solving analytical problems. Business Analytics consultants can package analytics solutions to specific problems as products offered. This book looks at Business Analytics consulting from this perspective. Some of the problems termed as products that we look at include customer loyalty (Chapter 7), customer and product segmentation (Chapter 8), customer profiling (Chapter 7), and so on. It is important to note that these problems, though common in retail business, also find expressions in other types of business.

It is important for Business Analytics consultants to know that data science techniques are embedded in business understanding. If you don't know how a business operates and how to improve a business normally, you cannot know how the data science technique will be used or where it fits in, to improve the sales of the business, for example. A good example is in using data science to improve ecommerce business sales. For example, to sell more of products not selling, you have to know how this can be done in typical marketing, for example, through cross-selling. This will not inform us of the kind of analytics to carry out. In this situation, you might want to use association rule analysis to discover products that have occurred together in the past and then use it for cross-selling.

A Business Analytics consultant might perform three types of roles when working with clients: expert (outside looking in, you're working with a client who has a problem and wants you to fix it), pair-of-hands (providing a service that the client company might be able to perform themselves,) and collaborator (working side by side, the outside consultant joins forces with members of the client organization to work on the project or solve the problem together). Each of these represents a different kind of interaction and a different source of satisfaction for the consultant.[3] When you take up consulting opportunities, it is important to state clearly which mode you want to operate with.

3.2 Managing Analytics Project

Typically, an analytics project is in phases; these phases will be covered in detail in Chapter 4. But this section will cover some important issues that relate to managing an analytics project successfully.

The first thing to do when you kick-start an analytics project is to establish the scope of the project, particularly stating clearly those things that are within the boundaries of the project. This should be made and documented at the beginning of the project. Clearly defining the goal of the analytics project will help to do this. This will help to take care of distractions and additions (even though valuable) that are not part of what you set out to do. If these additions are important, there can now be a basis for renegotiation.

The next and most important thing to put in place for managing an analytics project effectively is procedures for effective communication with all parties involved in the analytics project, from your team members to the stakeholders; they need to be kept in the know of what is going on whether good or bad, progressive or not. If need be, you need to also establish a line of escalation in case you need something resolved, for example, you are having issues retrieving data updates from the appropriate quarters. Closely related to this is the fact that you need to be able to monitor your team using whatever kind of tracker app you are comfortable with; this will enable you to get regular updates and identify and fix issues before it causes more damage to the analytics process. It will also help the communication of the analytics team. Finally, since the whole essence of consulting is to make money by solving business problems with analytics, then, there is a need to track hours and finances invested in the project so you know how to bill accurately, though this also depends on the billing approach you have chosen.

In addition to the above, for you to be able to successfully manage an analytics project you need to be able to conceptualize analytical solutions for real-world problems. This means that when the problem is presented to you in the form of a real-world or real business problem, you need to be able to translate it to an analytics problem (note that not all problems are solvable with analytics) and then break the solution down into smaller components and link them together to give the overall results as revealed in Figure 3-2.

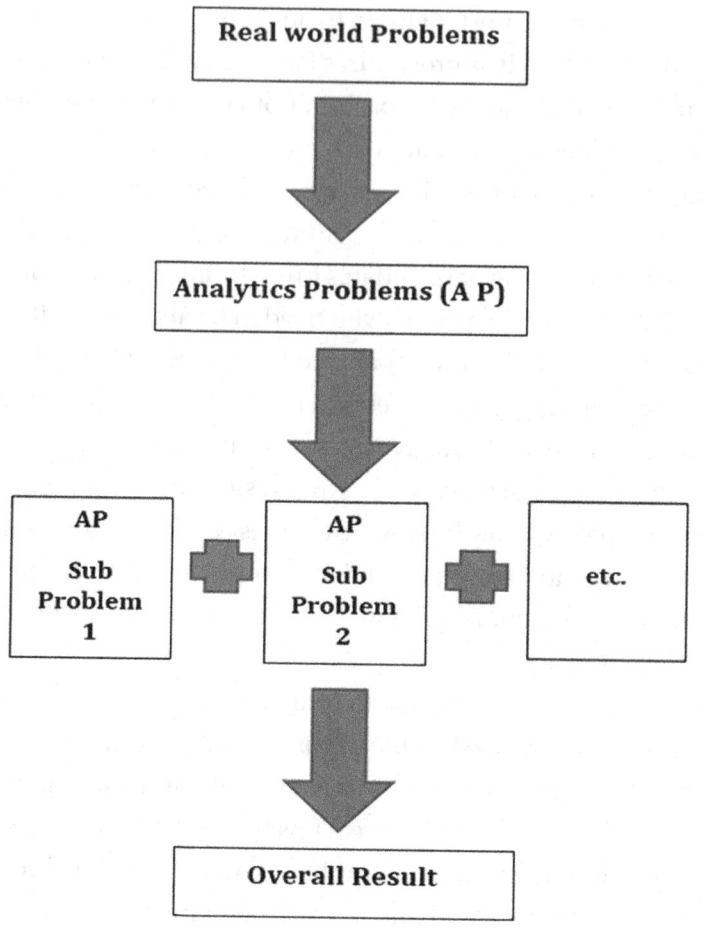

Figure 3-2. *Analytics problem solving process*

It is important to state at this point that machine algorithms and so on are not enough to solve business problems analytics-wise; there has to be a marriage of the business domain expertise (marketing, sales, etc.), technical expertise (data engineers, information technologist, etc.), and the technical skills involved in data analytics.

No matter the size of the business you are consulting for, there is a need to factor all the above into your analytics project to make it successful.

3.3 Success Metrics in Analytics Project

It is important to determine accurately the metrics to be used in evaluating the success of your analytics project because this helps to determine if the project has been successful or not. Not only that, it will help to bill your clients. As consultants, there is a need to leverage on previous success stories to bid for other contracts; therefore, getting accurate success metrics will make this possible. Even though there are general success metrics, the most effective metrics used in measuring the success of an analytics project is determined by the application area of that particular project. For example, if you are consulting for an ecommerce business with a goal to improve the sales or revenue, the metric to measure for success will include the AOV (average order value). In a physical retail shop, for example, the total number of customers that has been added due to implementing the results of the analytics project will be a good metric if the goal is to increase the customer base.

Generally, for a data science project, we have the following metrics:

- *Traditional metrics*: These types of metrics will help us to know how we are performing relative to the plan. This is where you want to check the time, budget, and scope of what is meant to be achieved.

- *Agile metrics*: These types of metrics help us to check at frequent intervals if we are providing value or not. These can be achieved by including cycle times and velocity/speed metrics.

- *Financial metrics*: These border around if the organization is truly increasing its financial value due to the results obtained from the analytics project. It is important to note that this might not be assessed only on completion but even after relevant milestones. Metrics such as these include revenue and cost metrics, return on investments, and so on.

- *Organizational goals*: You want to constantly check if your project is impacting the organizational goals and meeting the satisfaction of the stakeholders.

- *Software metrics*: In situations where the analytics project translates into software, you want to be sure if the quality of the software meets industry standards based on defect count, latency, and so on.

- *Model performance metrics*: Depending on the type of models used for your analytics project, you need to evaluate them based on the appropriate evaluation technique, for example, using the root mean square error (RMSE), precision and recall, and so on.

3.4 Billing the Analytics Project

First of all, it is important to say here that the billing in this section is talking from the perspective of a consultant outside the organization. Before billing any analytics consulting project, the very first thing to do is the initial analysis. The details of the initial analysis are covered in Chapter 4, Section 4.1. The initial analysis may or may not be billed depending on the state of your proposal. Sometimes, you might want to do an initial analysis for free to let the clients see the potential in buying into the analytics projects. If the situation is such that the client is already onboard and the initial analysis is such that it will involve some investment in time and resources as there is a need to troubleshoot some data-related issues, the initial analysis can be billed. The initial analysis is also billed in situations where it's the client that called for the analysis. What the initial analysis helps to do in billing the analytics project includes the following:

- First, it can be discovered, particularly with small businesses, that the data needed for the analytics project is not available; hence, the project cannot be carried out. Sometimes, after initial analysis, the best that can happen is to advise the clients on data governance procedures and revisit the project some other time.

- Second, the initial analysis will help you to get the realistic goals of the analytics project and the metrics to be used to evaluate the success. This will help you know what the components of the billing will be. If, for example, the goal is to increase the number of customers, and the return on investment for a customer is known, you can have an agreement on what percentage to collect on each increase in customers. This method usually works when the client is not willing to put a downpayment on the analytics project, and you still want to go ahead because you see a potential success story (this is usually the case with really small businesses).

- Third, the result of the initial analysis will help you to know the tasks involved in solving the problem and the amount of investment that will go into the project in terms of time, financial resources, and so on. All these have to be costed in detail and factored into the billing process. At this point, it is important to say that there might be a need to either have a business negotiator on your team or be educated in the art of negotiating. There are several online courses that could help here.

Data science consultants often offer their analytics service as a product. The billing can then be done just as a normal product is billed. Product-based billing can be done by the client paying for what they got out of the service rendered. So if you provide your analytics service and the metrics show success, you are paid based on the agreed amount. To bargain with the future result of the analytics project, you need to be able to clearly define and measure what the analytics project will achieve. So in essence, clients can be charged for an analytics project based on percentage increase in success criteria. Note that there might be a need to make the agreement legal. Providing your analytics service as a product could also be putting it in stages of completion, that is, kickoff, update, milestones, and so on. These stages can be pitched as a product and billed.

Also, an important warning is to beware of hidden costs of consulting, for example, when calculating the hours spent on the consulting project, some times spent on the analytics project might not be direct, for example, the time spent keeping up with the technology team, research, prospecting, accounting, proposals, marketing, internal trainings, code review, and so on. If you had to organize training, research and development, and so on, the time to do all these must be included in the cost.

Sometimes, when consulting for a small business, you might even need to do a first implementation of recommendation, and when they see the results, you now can charge for continuous implementation.

Finally, defining the scope of an analytics project is useful even in billing the clients as this will help to know where your analytics task starts and stops. Sometimes, when dealing with small businesses and because of the urge to get successful results, you might be tempted to go outside the typical boundaries of analytics, for example, gathering data, marketing, and so on, just to get your intended results. Depending on the situation at hand, this may or may not be part of the task, but it is important to state here that if it is, it should be factored into the billing both in terms of resources used and time spent.

3.5 References

1. Isabelle Flückiger (April 5, 2021) Discover 9
    Consultancy Segments to Start an Exciting
    Data Science Journey for Any Experience Level.
    https://towardsdatascience.com/discover-9-
    consultancy-segments-to-start-an-exciting-
    data-science-journey-for-any-experience-
    level-a972cb6b63e4

2. Modernanalyst.com (Assessed March 2022) Three
    Modes of Business Analysis Consulting.

3. Karl Wiegers (2019) Successful Business Analysis
    Consulting: Strategies and Tips for Going It Alone
    (Business Analysis Professional Development) None
    Edition published by J.Ross Publishing.

More resources on the chapter for further reading

- Joe McFarren (May 28, 20215) Tips for Managing a
  Successful Analytics Project. www.tessellationtech.
  io/5-tips-for-managing-a-successful-
  analytics-project/

- Data Analyst Career Switch | How I Became a Data
  Analyst, www.youtube.com/watch?app=desktop&v=o
  ZbEvdn9pbE

- Garrett Eichhorn (May 11, 2020) A Data
  Scientist's Guide to the Side Hustle. https://
  towardsdatascience.com/a-data-scientists-guide-
  to-the-side-hustle-3dd93a554eb8

- Brandenburg, Laura (2015) How to Start a Business Analyst Career: The handbook to apply business analysis techniques, select requirements training, and explore job roles leading ... career. Clear Spring Business Analysis LLC.

- Vicki James (2019) Leveraging Business Analysis for Project Success published by Business Expert Press.

CHAPTER 4

Business Analytics Consulting Phases

This chapter will look at the stages involved in Business Analytics consulting, mainly when the analytics service is offered as a product from either within or outside the business. We will look at the proposal and initial stages of analysis that gives direction to the analytics project. Then we look at the details involved in the pre-engagement, engagement, and post-engagement phases. It is important to know that the stages are presented in a typical or generic way, but when implemented, there might be a reason to modify or customize them for the application scenario.

4.1 Proposal and Initial Analysis

Before coming up with a proposal pitch, which will probably be the first time you interact with the organization, you have to do a research on the organization or SME to get some background information that will help your analytics proposal. In the proposal, the background information will help you to know and highlight the problems you plan to solve for the organization. Though this is not compulsory, it could help grab the attention of the stakeholders and make them interested in investing in the analytics project. The major hindrance to jump-starting the analytics project with SMEs is that the stakeholders don't see a need or an urgency

© Afolabi Ibukun Tolulope 2022
A. I. Tolulope, *Data Science and Analytics for SMEs*,
https://doi.org/10.1007/978-1-4842-8670-8_4

to invest in such a project at the moment given the fact that they are still making some income based on their present methodology. The onus now lies on the consultant to convince them on the transformation and achievement that data analytics could bring about. In situations where this stage is not needed (i.e., the stakeholders are already onboard), you can just have a brief introduction meeting which will detail the analytics procedure and elicit the kind of problems that are relevant to the business. It will also help at this point to get feedback from the stakeholders and know what is on their mind and what they hope to achieve with the analytics project. Though there are situations where the stakeholders might not know what constitutes analytics problems, they will definitely know the needs of the organization. At this stage, very good presentation skills will be needed to get your proposal accepted. At the proposal stage, it will help to present problems that have been customized for the business in particular. The consultant needs to be careful at this stage not to give false hope or to propose to solve problems for which data resources are not available, all in a bid to get the proposal accepted.

In the initial analysis stage, you want to find out what is available in terms of the data, data quality, and integrity. This is the stage where you explore the proposed problems with respect to the data to be used in solving each of the problems. At the end of this phase, you should be able to give a detailed report of what is achievable given the available data. Also, in the initial analytics stage, you need to understand the organization's operational procedure (conceptual business model) as explained under the Business Analytics journey in Section 1.3 of Chapter 1. This will enable you to be able to understand the optimization approaches for the proposed analytics models and how the result of your analytics will fit into the organization or business process. You will also be able to discover when and who will be responsible for implementing your recommendation and so on. With this, you are able to know who is to be involved in the communication stages of the analytics project. After the initial analysis, you should also be able to give an idea of what are the expected outcomes of the analytics project at a high level of abstraction and advice on best data practices for the organization. The tasks involved in the initial analysis stage can be summarized as follows:

- Discovery and assessment of the available data (internal and external) for analytics and the quality.

- Highlight the details on how to access the data available.

- Understand the organizational database and stage the process of integrating the data for analysis if need be.

- Discover the analytics obtainable from the current data.

- Highlight the decision objectives and determine the analytics output that will help to make an intended decision feasible, given the available data.

The output of the initial analysis stage includes

- Documentation on the detailed discoveries as regards the current data available and obtainable and an abstraction of what benefits can be obtained from it

- Advice on the data platform upgrades and integration, particularly toward future analytics

- Specific highlights of the objectives of the analytics project, decisions that can be made, and the analytics output that will help to make such decisions feasible given the available data

- Projects that will be carried out, stating the milestones, deliverables, costs, and expected ROIs

At this stage, you need to make it clear that the analytics project is an iterative process, which means that you might need to make recommendations which will be implemented and evaluated and, if necessary, repeated. This initial analysis will help to

- Learn what is possible with your data

- Have a fresh set of eyes on your current data pipelines, warehouses, and analytics strategy

- Give advice on data platform upgrades

In the initial analysis stage, you want to understand whatever data is available, that is, inquiring with subject matter experts about what field names might mean and seeking internal tribal knowledge of data source usage. There might even be a need to explore whatever data was given at this stage to know its feasibility to solve the problem because it is not uncommon to land in situations where the data you discovered from the inception is not as rich as you thought it would be for analysis.

Take note that during the initial analysis, you need to be able to determine how to avoid wasting time on unrealistic clients and let go of a project if it will not lead to a success story whether based on data resources or the negative business environmental situations. Finally, learn to navigate hype and unrealistic expectations. The rest of the analytics project is captured in three phases as presented in Figure 4-1. Just as mentioned earlier, this can be customized and expanded based on the nature of the individual analytics project. For the remaining part of this chapter, we will pick each of the stages one after the other and explain them in detail.

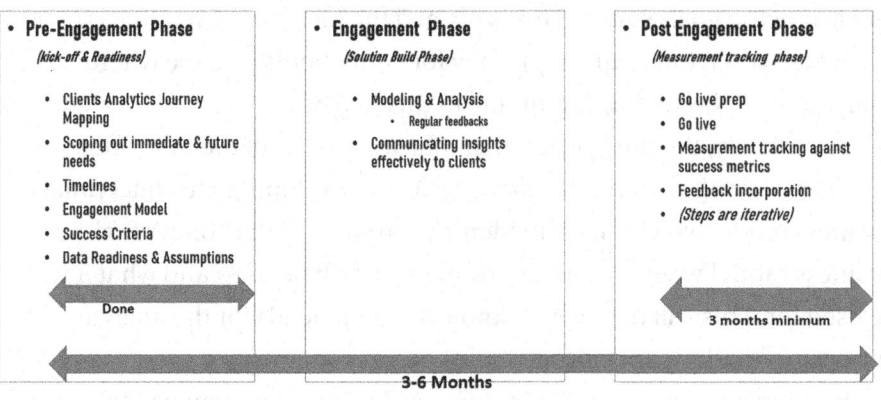

Figure 4-1. *Analytics consulting phases*

A typical analytics project consists of three major phases as revealed in Figure 4-1. The whole duration for the project varies based on the data situations and many other factors in the business domain, but, typically, it should cover an average of three to six months in which the post-engagement phase should be given at least three months. A good example of the IBM consulting process can be found at `https://365datascience.com/trending/ibm-data-science-consulting/`.

4.2 Pre-engagement Phase

In the pre-engagement phase, there is a need to carry out an exhaustive exploratory analysis using descriptive analytics to expose or discover the situation of things in relation to the intended goals. At the end of this phase, there might be a need to refine the goals based on their feasibility assessment. For each goal, you need to highlight the present situation using descriptive analytics, the conclusions that can be deduced, and the recommendations for immediate implementation or future analytics. There is a need to define success criteria, data readiness, and assumptions for each of the goals stated. This is termed the kickoff and readiness phase. It is where you try to map the journey for your clients and see where analytics can play an important role in this journey.

Technically, the pre-engagement phase consists of the following:

Clients' analytics journey mapping: After developing the conceptual business model, you then try to identify the stage of the conceptual business model where the analytics solution will be used and what it will be used for. This will help you to know the data needs for the analytics project, particularly specific attributes.

Scoping out immediate and future needs: To do this, you will take the goals of the analytics project and detail what you need to solve that is actually available and that needs to be available in the future. For example, if there is a need to clone the external market, what external data will be needed, and is it possible to get access to it?

Timelines: There is a need to state the time that each of the phases will take so that all parties will be carried along. This has to be done by stating clearly the milestones and time of their accomplishments.

Engagement model: This model will help to give all parties involved an idea of where they will be featuring in the project and what their role will be. It can also include a solution architecture that gives the clients an idea of how the analytics project will progress. Figure 4-2 is an example of a simple solution architecture together with the engagement model for consulting for an ecommerce business.[2]

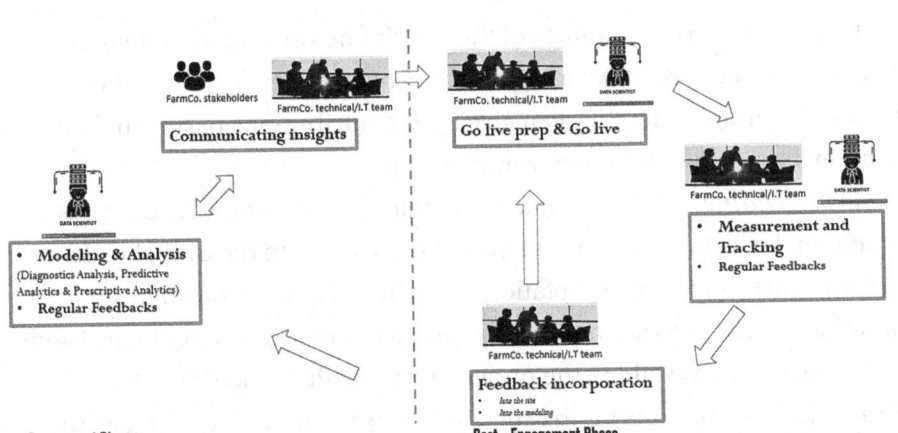

Figure 4-2. *Sample of a solution architecture and engagement model*

The solution architecture and engagement model presented in Figure 4-2 captures the pre-engagement and the post-engagement phase. This has been highly customized for a particular Business Analytics project directed at improving online sales. A different kind of business might interpret this in a slightly different manner based on contextual information. In the model presented in Figure 4-2, we can see some important characteristics; first, it is iterative, and there is a need to revisit some stages based on the result from the stages after it. Also, all parties that will be responsible for making the analytics project successful must be carried along, and their inputs are important to the success of the project. The solution architecture is very important because it helps to demonstrate the competence of the analytics team in solving the problem and helps the stakeholders to know what will be involved in getting the job done, and this is helpful particularly in getting whatever resources will be needed to execute the project.

Success criteria: For each of the goals, there is a need to use descriptive analytics to present the present situation in relation to the relevant success metrics as explained in Section 3.3 in Chapter 3. To do this, you are to give a detail of the values of the metrics before and after the goal is accomplished. The stakeholders should be communicated on these success criteria, so that they can give their contributions based on their domain knowledge and organizational goals.

Data readiness and assumptions: Finally, the pre-engagement phase should make it clearly known what is obtainable from the available data. This will clear any false expectation from the analytics project. In some situations, it could either redirect the project or cause some data decisions to be made such as stalling the project to curate data and so on. By stating the assumptions for the successful completion of the project, the stakeholders are able to know why some things do not go as planned.

At the end of the pre-engagement phase, there is a need to come up with the documents containing the details of the pre-engagement phase, which is presented to the stakeholders. This is important because oftentimes there would have been a further modification and refinement of the conclusions from the initial analysis stage.

4.3 Engagement Phase

Depending on the kind of analytics project, meaning, for example, we are expected to come up with software or recommendations for implementation or both, the engagement phase is when you build the solution to the problem or goal identified.

The engagement phase typically includes modeling and analysis, regular feedback, and communicating insights effectively to clients. The components of the engagement phase stated might increase depending on the project at hand but not reduced as each of these components determine the success of the project.

Modeling and analysis: First, determine the data mining task. Will it be descriptive analytics, predictive analytics, or prescriptive analytics? Will it be a combination of all these or will the task be used to complement each other? Sometimes, the predictive and prescriptive use descriptive techniques in the exploratory stage. This involves translating the general question or problem into a more specific statistical question.

Depending on the problem at hand, there are several ways to approach them; for example, the problem can be an anomaly detection problem where you identify unusual data records or association modeling, where you search for relationships between variables. It could even be clustering (discovering groups and structures in the data that are in some way or another "similar," without using known structures in the data), classification (generalizing known structures to apply to new data), regression (finding a function which models the data with the least error), or summarization (providing a more compact representation of the dataset, including visualization and report generation).[3]

In practice, you can select from descriptive, predictive, prescriptive, cognitive, or diagnosis. You can also perform descriptive first and, based on the results, use diagnosis to further investigate why and then use predictive and prescriptive analytics to solve the problem. You might even need to combine some or all of them. Most times, the result of the descriptive and diagnosis will determine the predictive technique to use.

The data mining technique to be used is determined by the size of the dataset, the types of patterns that exist in the data, whether the data meets some underlying assumptions of the method, how noisy the data are, and the particular goal of the analysis. It is therefore important to apply several different methods and select the one that is most useful for the goal at hand. It is important to note that the modeling process is iterative, and there is a need to try several variants of the technique and use multiple variants of the same algorithm. The final choice is determined by performance assessment of the technique or algorithm as the case may be.[1]

Feedback: In the engagement phase, there is a need for continuous feedback between the analytics team and the stakeholders, and this can be as many as the turn of events. Sometimes, this feedback could alter the course of action, encourage the domain experts to cooperate more, and so on.

Communication of insights: Insights will be communicated at regular intervals. Some insights can even cause some proposed future analytics to change and rework the whole analytics project. Communicating insights requires the knowledge of quality presentation skills orally and in presenting the results in the form of charts together with mastery in storytelling. In communicating insights, you should be able to gauge your audience and see what is appropriate for such an audience; sometimes, a high level of abstraction is needed, and sometimes there is a need for details. Results need to be presented using visuals where necessary, and finally tie up your recommendation with the insights that you have generated.

4.4 Post-Engagement Phase

After communicating the recommendations of your analytics project, the next phase is the post-engagement phase, also known as the measurement tracking phase. This phase consists of the go-live preparation, go-live, measurement tracking against success metrics, and feedback incorporation. The steps in the post-engagement phase are iterative.

Go-live preparation: This is when you organize workshops and training to educate those that will implement the recommendations and how they will implement them such that the overall goal of the analytics project can be achieved. Sometimes, there might be a need to use demos, fliers, and videos just to pass the message across. You also need to factor in change management, that is, how the people, processes, and systems that drive the organization will respond to the change that the recommendation will bring about to achieve the overall goal. If the solution involves a technology product, for example, you also need to tidy up loose ends as to how the solution will fit into the existing system.

Go-live: In this step, you deploy your solution, implement the recommendation, and supervise the implementation of the recommendation. It takes two to three months for the implementation to go on before the tracking begins.

Measurement tracking against success metrics: The next step is to track the effectiveness of your recommendations based on the success metrics highlighted in the pre-engagement phase. Based on the results from tracking the metrics, there are decisions to be made at this point. You might need to decide if there is a need to iterate the modeling processes or adjust the recommendations to improve success metrics or conclude on the achievement of the desired goal. Sometimes, you might need to track compliance techniques; for example, if it's a recommender system that has recommended some products, you want to find out if those products ended up being sold or not.

Feedback incorporation: The feedback from the go-live, measuring, and tracking will cause some actions to be taken toward the success of the project, particularly in improving the analytics process.

Finally, when the measurement reveals a level of success that is acceptable, the next thing is to see how the process can be automated to avoid heavy dependency on the manual implementation of recommendations in the future. Though this is optional, having things run themselves is more efficient than relying on a manual process if it can be afforded.

4.5 Problems

Given the following scenarios, explain the details of the activities for the pre-engagement, engagement, and post-engagement phases of a freelance business analyst that was given a contract based on the following:

> a. A textile retailer, though having been experiencing success early, is now starting to see a decline in sales. Thankfully, the retailer has a dashboard that

collects demographic information about current and target customers. The goal of this retailer is to use this information to locate areas for improvement and identify where sales were strongest. One way to go about this is to segment buyers by relevant factors and customize marketing strategies to each group.

b. A fitness instructor noticed that customers were leaving his business. The intention is to develop and install a predictive analytics model that will identify if a customer is likely to leave and quickly offer incentives that could prevent them from leaving.

4.6 References

1. Ramesh Sharda, Dursun Delen, Efraim Turban (2020) Analytics, Data Science & Artificial Intelligence Systems for Decision Support. Eleventh edition, published by Pearson.

2. Vishranth Chandrashekar (July 2020) Journey Into Data Science Consulting | How To Become A Data Science Consultant | Great Learning, www.youtube.com/watch?v=Rnhv_gayPHs

3. Jiawei Han, Micheline Kamber, and Jian Pei (2012) Data Mining Concepts and Techniques, Third Edition published by Elsevier.

CHAPTER 5

Descriptive Analytics Tools

This chapter is focused on the mostly common descriptive analytics tools used in business generally and specifically in small businesses. The chapter will help to use descriptive analytics tools to understand your business and make recommendations that can improve your business profits. For small businesses, descriptive analytics helps SMEs to make sense of available data to monitor business indicators at a glance, help SME owners observe sales trends and patterns on an overall basis, and deep dive into product categories and customer groups. It also helps SMEs plan product strategy, pricing policies that will maximize their projected revenues and derive valuable insights for getting more customers.

5.1 Introduction

In Business Analytics, descriptive analytics helps to link the market to the firm through information and provide information needed for actionable decisions. Therefore, descriptive analytics can be seen as the principles for systematically collecting and interpreting data that can aid decision makers. There are three types of descriptive analytics:

© Afolabi Ibukun Tolulope 2022
A. I. Tolulope, *Data Science and Analytics for SMEs*,
https://doi.org/10.1007/978-1-4842-8670-8_5

Exploratory research: This is used in situations where there is a need to solve an ambiguous problem, for example, to conduct exploratory research to find out why sales are declining.

Descriptive research: It provides answers to questions bothering on awareness, for example, what kind of people are buying our goods?

Causal research: This is used in situations where the problem is clearly defined, for example, will buyers purchase more of our products with a change of our store arrangement?

The rest of this chapter picks each of the common descriptive analytics tools and gives detailed explanation on their uses and a practical demonstration of how to use them. The data used for the practical demonstrations in this chapter is named SuperstoreNigeria.xls.

RapidMiner can be used for several types of visualization, but we have selected the popular and most common ones in this chapter.

5.2 Bar Chart

The bar chart can be used to examine categorical data for unexpected patterns of results, for example, there may be many more of some categories than others. Some categories may be missing completely or there might even be uneven distributions in the categories. There might even be extra categories, for example, if there was a mistake and the gender was recorded as "M" and "F," but also as "m" and "f," "male" and "female." There may also be unbalanced experiments such that some data are missing or unusable; this can be discovered using bar charts. Bar charts can also help to discover when we have too many categories and make decisions as to whether we are to combine them and how. Finally, it can reveal errors and missing values.

Creating a bar chart in RapidMiner

1. You can create a new repository or use an existing one and ***Import*** the data named *SuperstoreNigeria.xls* (refer to Section 2.4 in Chapter 2 for a guide).

2. Figure 5-1 is a screenshot of the imported data in RapidMiner. Click Next and select the repository where you want to put the data.

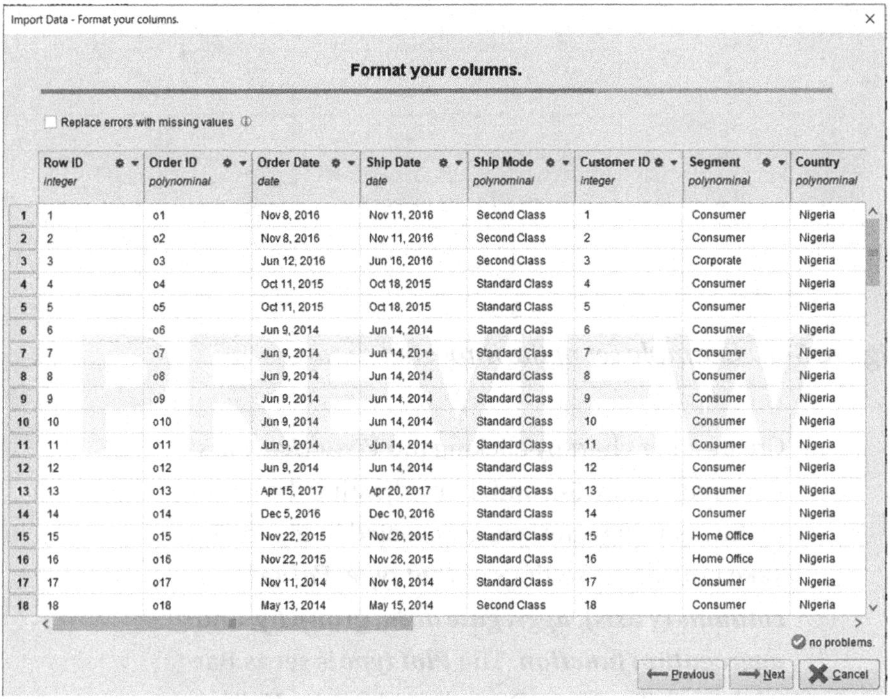

	Row ID (integer)	Order ID (polynominal)	Order Date (date)	Ship Date (date)	Ship Mode (polynominal)	Customer ID (integer)	Segment (polynominal)	Country (polynominal)
1	1	o1	Nov 8, 2016	Nov 11, 2016	Second Class	1	Consumer	Nigeria
2	2	o2	Nov 8, 2016	Nov 11, 2016	Second Class	2	Consumer	Nigeria
3	3	o3	Jun 12, 2016	Jun 16, 2016	Second Class	3	Corporate	Nigeria
4	4	o4	Oct 11, 2015	Oct 18, 2015	Standard Class	4	Consumer	Nigeria
5	5	o5	Oct 11, 2015	Oct 18, 2015	Standard Class	5	Consumer	Nigeria
6	6	o6	Jun 9, 2014	Jun 14, 2014	Standard Class	6	Consumer	Nigeria
7	7	o7	Jun 9, 2014	Jun 14, 2014	Standard Class	7	Consumer	Nigeria
8	8	o8	Jun 9, 2014	Jun 14, 2014	Standard Class	8	Consumer	Nigeria
9	9	o9	Jun 9, 2014	Jun 14, 2014	Standard Class	9	Consumer	Nigeria
10	10	o10	Jun 9, 2014	Jun 14, 2014	Standard Class	10	Consumer	Nigeria
11	11	o11	Jun 9, 2014	Jun 14, 2014	Standard Class	11	Consumer	Nigeria
12	12	o12	Jun 9, 2014	Jun 14, 2014	Standard Class	12	Consumer	Nigeria
13	13	o13	Apr 15, 2017	Apr 20, 2017	Standard Class	13	Consumer	Nigeria
14	14	o14	Dec 5, 2016	Dec 10, 2016	Standard Class	14	Consumer	Nigeria
15	15	o15	Nov 22, 2015	Nov 26, 2015	Standard Class	15	Home Office	Nigeria
16	16	o16	Nov 22, 2015	Nov 26, 2015	Standard Class	16	Home Office	Nigeria
17	17	o17	Nov 11, 2014	Nov 18, 2014	Standard Class	17	Consumer	Nigeria
18	18	o18	May 13, 2014	May 15, 2014	Second Class	18	Consumer	Nigeria

Figure 5-1. *Screenshot of the imported data*

3. On the **Results** interface in Figure 5-2, you can
    click the **Data**, **Statistics**, **Visualizations**, and
    Annotations to explore the various characteristics
    of the data as explained in Chapter 2, Section 2.4.

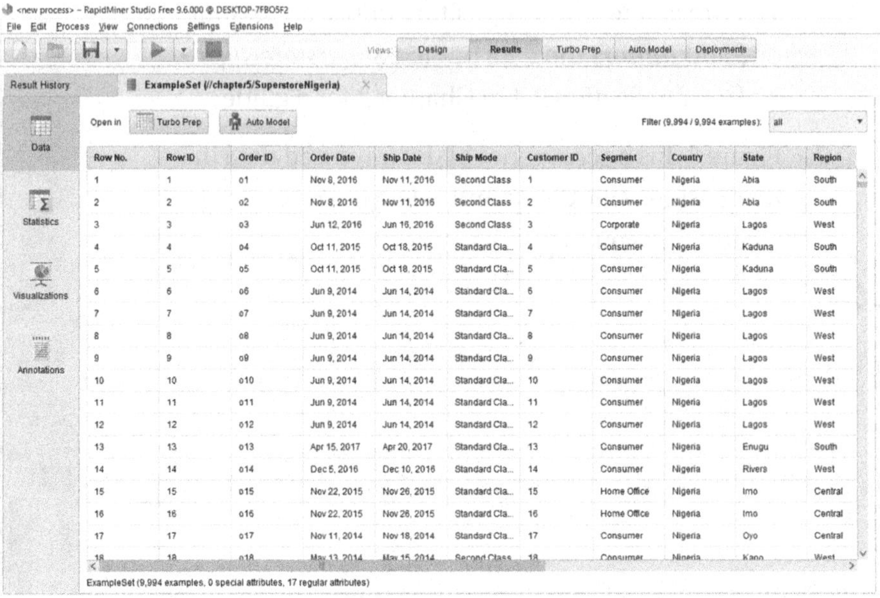

Figure 5-2. *Example set of the imported data*

4. Create a bar chart by clicking the **Visualizations**
    button in Figure 5-2 and then adjusting the setting
    of the resulting interface as revealed in Figure 5-3.
    Take note of the settings of **Plot type**, **Value
    columns** (y axis), **aggregate data**, **group by**, and
    aggregation function. The **Plot type** is set as Bar
    (Column). The Row ID is selected for the **Value
    columns** so as to count each row in the data to
    create the bar chart. **Aggregate data** is selected to
    indicate we will be using aggregated calculations.

The **group by** is selected indicating the categorical variable (segment) we want to examine. The **aggregation function** indicates that we want to know how many rows exist for each category in the categorical attribute called segment.

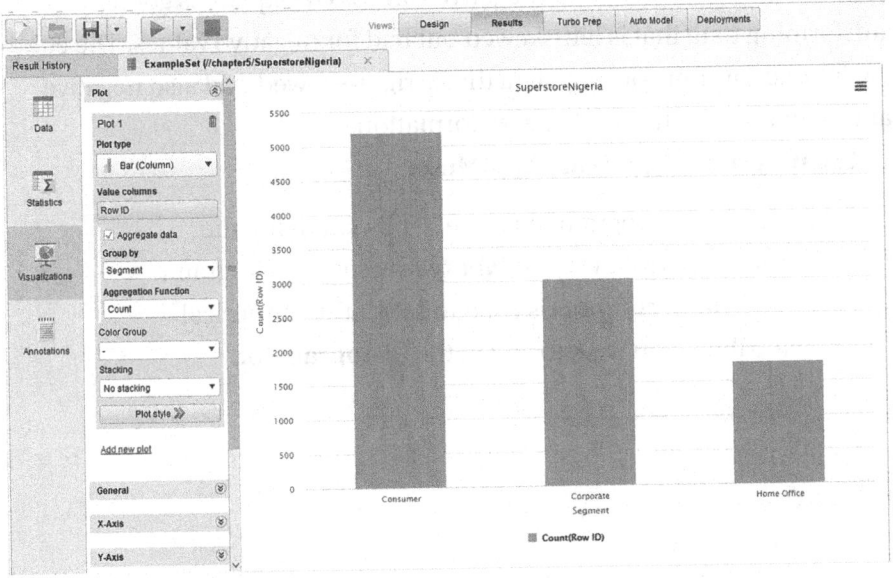

Figure 5-3. *Bar chart example*

5.3 Histogram

A histogram is used to group data into intervals, and drawing a bar for each interval shows the empirical distribution. Histograms help to understand the center of the data. Histograms are an excellent tool for identifying the shape of your data distribution. There are situations where data is either skewed to the right or left, and you have to perform data transformation. When using a histogram to examine your data for prediction, for example, you want to make sure the data is not skewed to the right or to the left in such a way that it could affect the result of your analysis. That is, the data needs to be symmetric or almost symmetric.

The log method is good for transforming right-skewed data and bad for zero and negative values. The square root method is good for transforming right-skewed data and bad for negative values. The square is good for transforming data that is left-skewed but bad for negative values. The cube root is good for transforming data that is right-skewed and also negative values but not effective for log transformation.

Creating a histogram in RapidMiner

1. Create a histogram in Figure 5-4 by changing the settings as follows; the *Plot type* is set as Histogram. The **value** *columns* is used to select the numerical attribute you wish to create the **histogram** for.

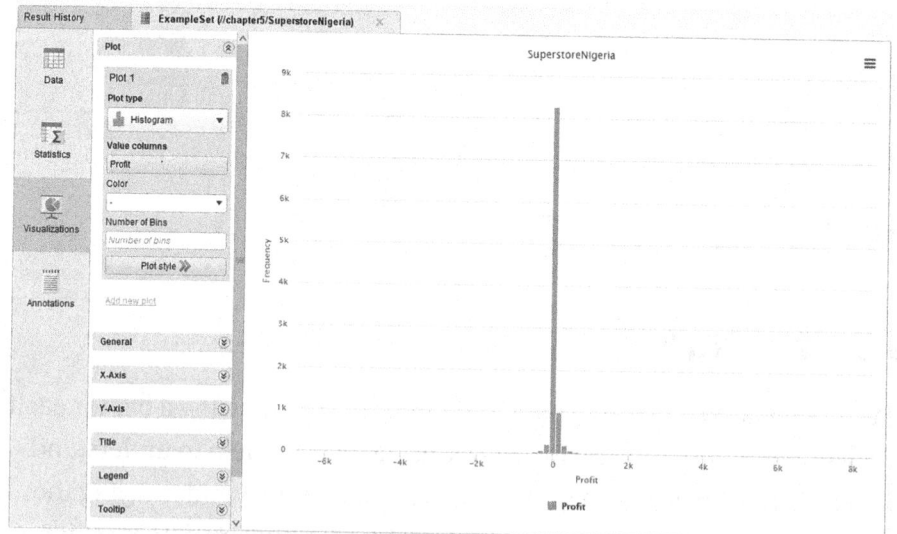

Figure 5-4. *Histogram of the attribute profit*

2. Figure 5-5 is the histogram of the attribute quantity.

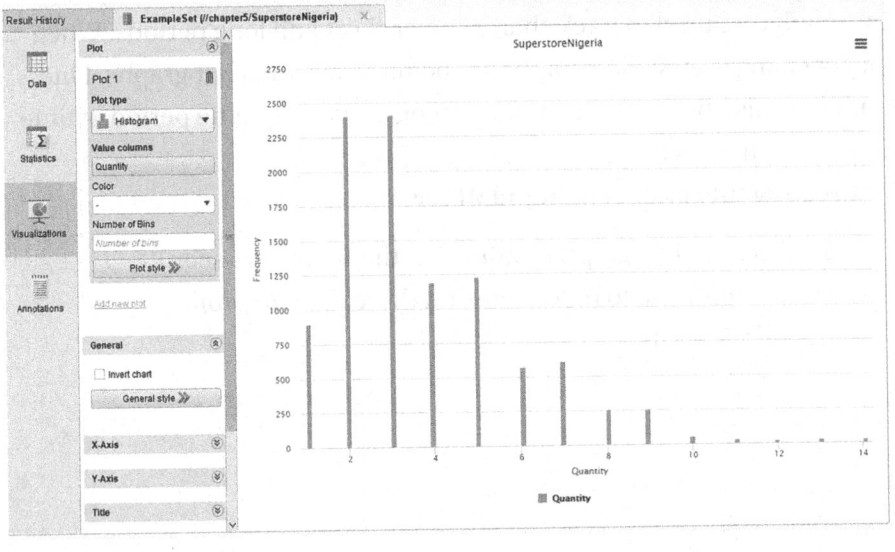

Figure 5-5. *Histogram of the attribute quantity*

3. The histogram in Figure 5-4 is fairly symmetric,
    while that in Figure 5-5 is slightly skewed to the
    right. A histogram that is symmetric has data evenly
    distributed on both sides of the line that divide the
    histogram into two equal halves vertically. The value
    on the x axis where this happens on the histogram
    plot is the median, which is also equal to the mean
    and equal to the mode (for symmetric histogram
    shape). We can also use the histogram to pick where
    most of the values fall between.

5.4 Line Graphs

Line graphs are used to track changes over short and long periods. When smaller changes exist, line graphs are better to use than bar graphs. Line graphs can also be used to compare changes over the same period of time for more than one group.

Creating line graphs in RapidMiner

1. Create a line graph by adjusting the settings as seen in Figure 5-6 to reflect *sales* on the x axis and *profit* on the y axis.

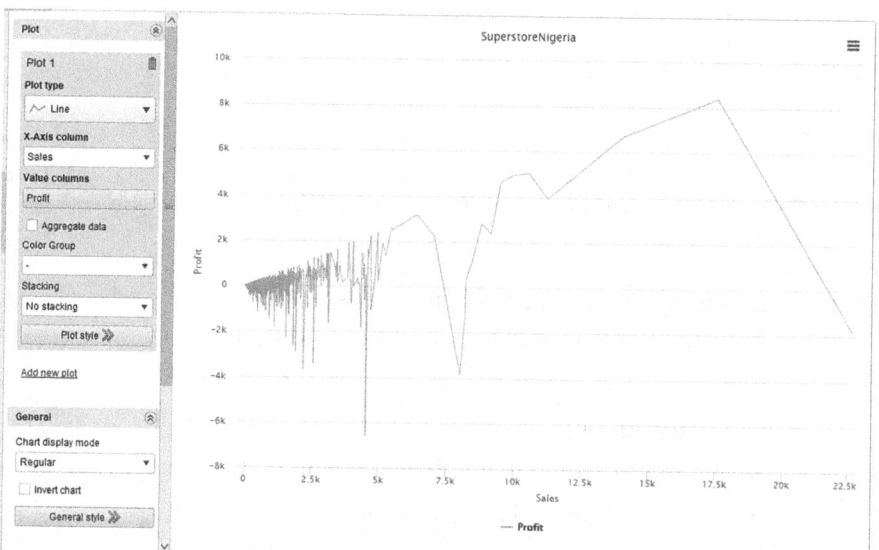

Figure 5-6. *Example of a line graph*

2. From Figure 5-6, we can see how profit (y axis) changes with sales (x axis).

5.5 Boxplots

A boxplot is a standardized way of displaying the distribution of data based on a five-number summary: minimum, first quartile (Q1), median, third quartile (Q3), and maximum. It can tell you about your outliers and what their values are. It can also tell you if your data is symmetrical, how tightly your data is grouped, and if and how your data is skewed. It can also be used for comparisons of distributions across subgroups.

Creating boxplots in RapidMiner

1. Create a boxplot for the attribute quantity by adjusting the settings as seen in Figure 5-7. The **Plot type** is set to Boxplot, while the **Value columns** (y axis) is set to quantity.

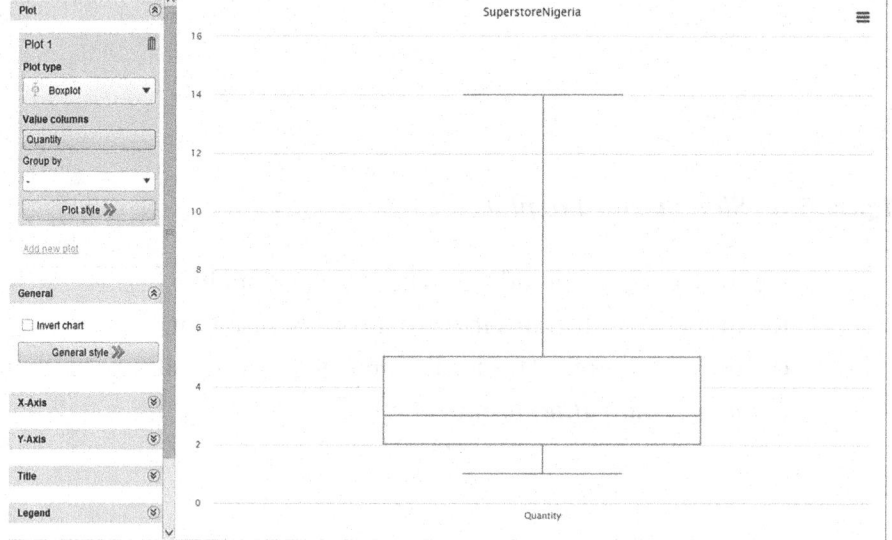

Figure 5-7. *Example of boxplots*

2. From Figure 5-7, the minimum quantity is 1, the first quartile (Q1) is 2, the median is 3, the third quartile (Q3) is 5, and the maximum is 14. The majority of the quantity falls between 2 and 5.

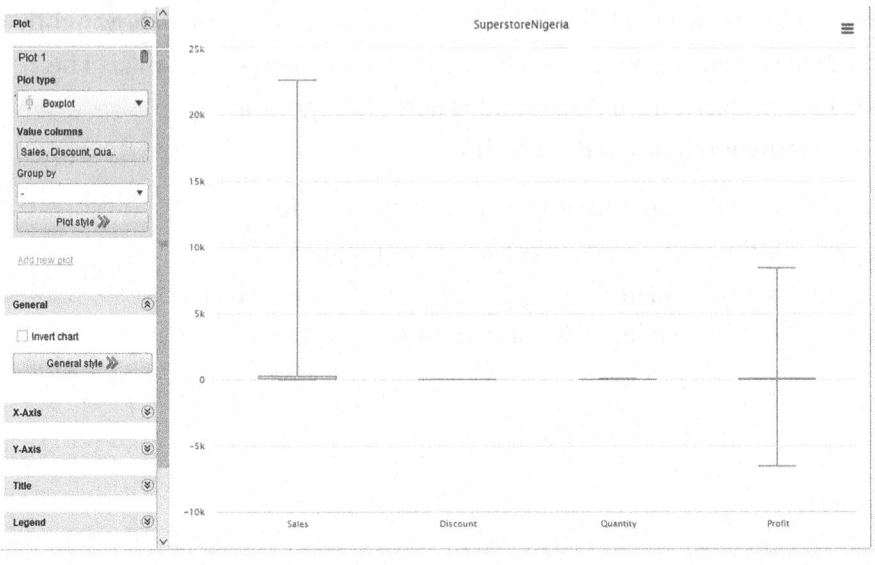

Figure 5-8. *Side-by-side boxplots*

3. Figure 5-8 shows the side-by-side boxplots for all the numerical attributes in the data: sales, discount, quantity, and profit. This is obtained by including all of them in the ***Value columns***.

5.6 Scatter Plots

The scatter plots help to discover features such as

- *Causal relationships (linear and nonlinear)*: One variable may have a direct influence on another in some way. For example, people with more experience tend to get paid more. It is standard to put the dependent variable on the vertical axis.

- *Associations*: Variables may be associated with one another without being directly causally related. Children get good marks in English and in Math because they are intelligent, not because the ability in one subject is the reason for the ability in the other.

- *Outliers or groups of outliers*: Cases can be outliers when the scatter plot is created for two dimensions (one on the x axis and the other on the y axis), but not outliers in either dimension separately. Taller people are generally heavier, but there may be people of moderate height who are so heavy or light for their height that they stand out in comparison with the rest of the population.

- *Clusters*: Sometimes, there are groups of cases that are separate from the rest of the data.[1]

Creating scatter plots in RapidMiner

1. Create a scatter plot of sales on the x axis and quantity on the y axis by adjusting the settings as seen in Figure 5-9. The ***Plot type*** is set to scatter plot.

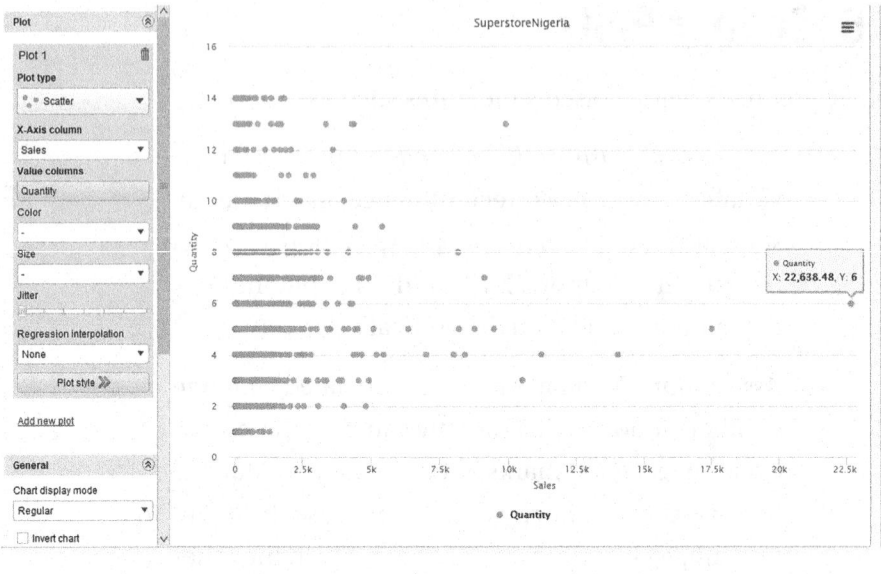

Figure 5-9. *Scatter plot*

2. A color-coded scatter plot can be created by adding
    a categorical variable to the setting (**color**) in
    Figure 5-9 to give Figure 5-10. The color-coded
    scatter plot helps to further explain the relationship
    between attribute sales and quantity using a
    categorical variable (category to separate that
    relationship). From Figure 5-10, we can see that
    items with low quantity and low sales mostly belong
    to the technology category.

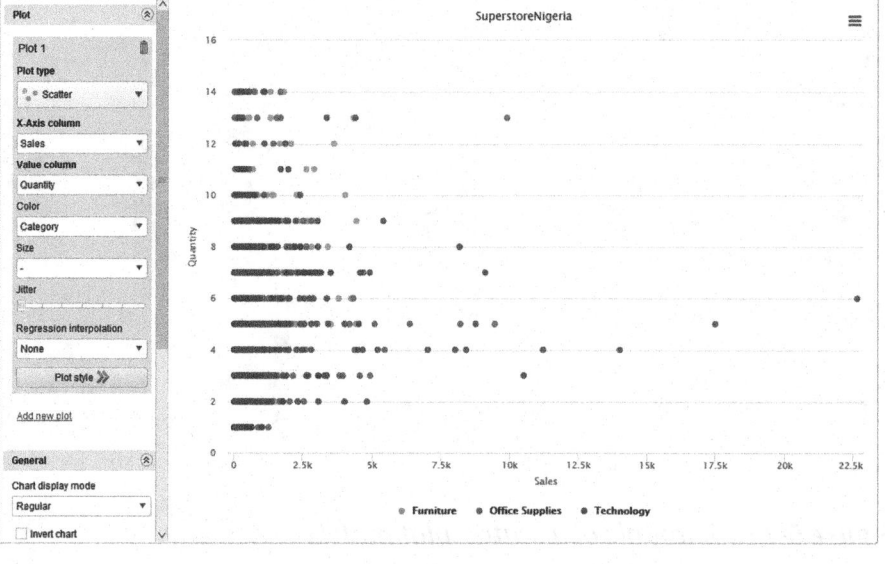

Figure 5-10. *Example of color-coded scatter plot*

3. Let us now create a scatter plot matrix for the
    numerical attributes discount, quantity, and sales
    using the settings in Figure 5-11. The Plot type is set
    to Scatter Matrix; Value columns are set to Sales,
    Quantity, and Discount; and Column Summary is
    set to Histogram. The scatter plot matrix will help to
    compare all the attributes intended at once.

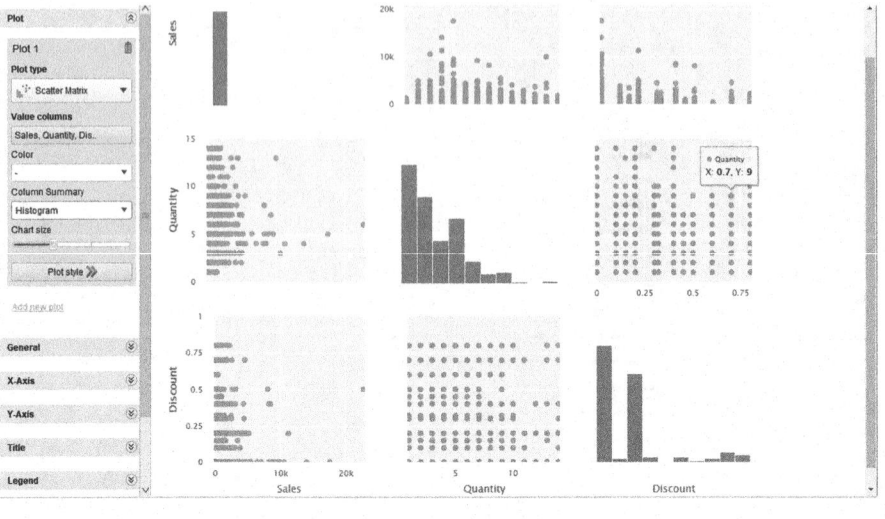

Figure 5-11. *Example of a scatter plot matrix*

5.7 Packed Bubble Charts

Bubble charts display data as a cluster of circles. It can be used to visualize the size of categorical attributes. Each category of the categorical attribute represents a circle, whereas the number in each category represents the size of those circles.

Creating packed bubble charts in RapidMiner

1. To create bubble charts, the data named *OnlineQuestion1.xls* will be used. **Import** the data into your repository; in the design view, link the data to the result (***res***), then run the process, and on the resulting interface (i.e., ***Results***), click the visualizations tab. (Note that the configuration of the free RapidMiner edition used for this course does not allow for visualizing more than 2000 rows for packed bubble charts.)

2. Create the bubble charts using the setting as shown in Figure 5-12. The **Value column** (y axis) is set to Age. The **Category** is set to sex (the categorical attribute under investigation).

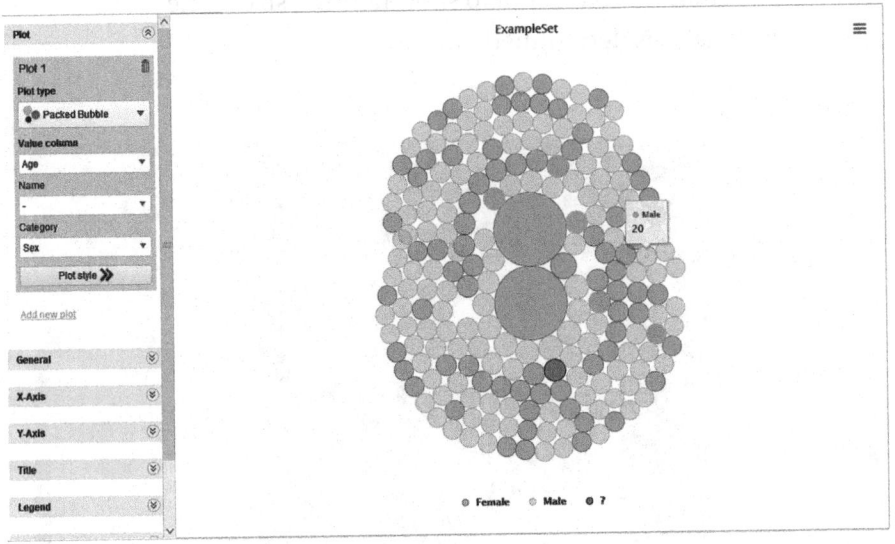

Figure 5-12. *Packed bubble chart*

5.8 Treemaps

Treemaps display data in nested rectangles. It can be used to visualize the size of categorical attributes. Each category of the categorical attribute represents a rectangle, whereas the number in each category represents the size of those rectangles.

Creating treemaps in RapidMiner

1. To create treemaps, the data named *OnlineQuestion1.xls* will be used. (Note that the configuration of the free RapidMiner edition used for this course does not allow for visualizing more than 2000 rows for treemaps.)

2. Create the treemap using the setting as shown in
    Figure 5-13. The **Value column** (y axis) is set to
    Age, **Group by** is set to LikeMaths (the categorical
    attribute under investigation), and the **Aggregation
    Function** is set to count to show how the size of the
    rectangles is determined.

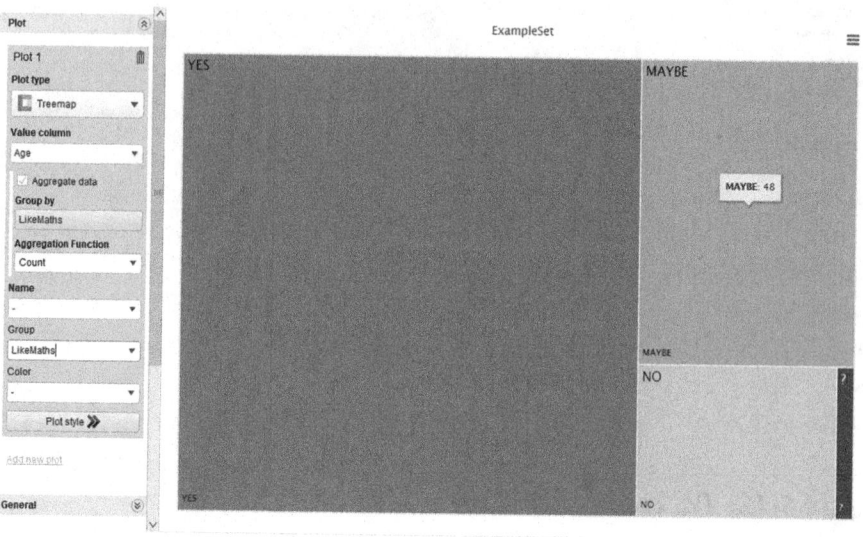

Figure 5-13. *Treemaps*

3. From Figure 5-13, Yes has the highest count, and
    this is indicated by the largest rectangle color blue.
    This is followed by Maybe and so on.

5.9 Heat Maps

A heat map is a graphical display of numerical data where color is used
to denote values. They are especially useful for the purpose of visualizing
correlation tables and visualizing missing values in the data. To do this,
the information is conveyed in a two-dimensional table. A correlation

table for a set of attributes has the same number of rows and columns. If the number of attributes is a lot, then a subset of the attributes can be used. Heat maps make it easy and faster to scan correlation values with the color codes.

Creating heat maps in RapidMiner

1. To create heat maps for visualizing missing data, for example, we will use the data named *OnlineQuestion1.xls* and make the settings as displayed in Figure 5-14. The **Plot type** is heat map, the attributes selected are Age, AmountToPay, Message, and Moved_AD. We will look at heat maps for correlation in Chapter 6.

2. In Figure 5-14, given the selected attributes, the whole rows (223) in the data are displayed. The missing data for a particular attribute is represented in color white. Visualizing the data like this will help to have an idea of how much data is missing. For example, there are very few cells with missing data.

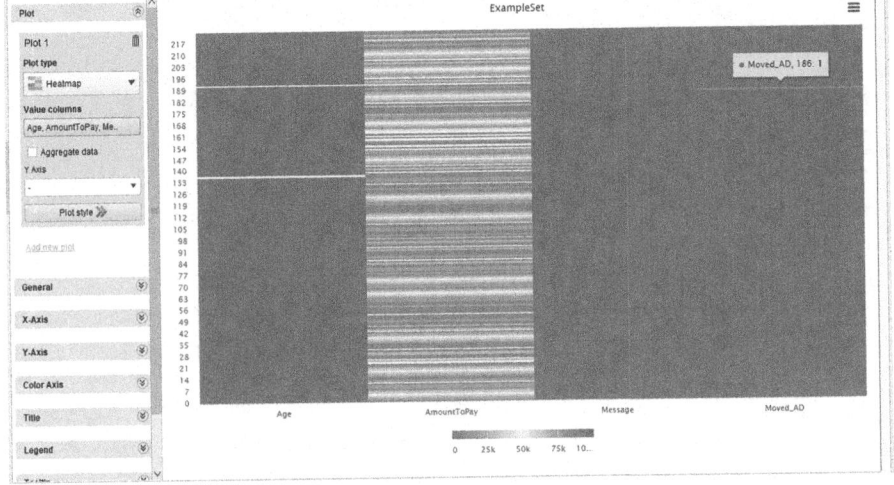

Figure 5-14. *Heat map for missing data*

5.10 Geographical Maps

Maps are basically the geographical representation of your data. There are certain types of inferences that are best achieved only when the data is visualized geographically.

Creating geographical maps in RapidMiner

1. For the purpose of demonstration, we will use the *customersNigeria.xls* data. This data consists of the customers of a particular business and their location data.

2. To create the geographical maps, import the data into your repository; in the design view, link the data to the result (***res***), then run the process, and on the resulting interface (***Results***), click the visualizations tab.

3. Create the setting as shown in Figure 5-15. The ***Plot type*** is point map; ***Select map*** is Nigeria. ***Latitude*** is latitude, and ***Longitude*** is longitude.

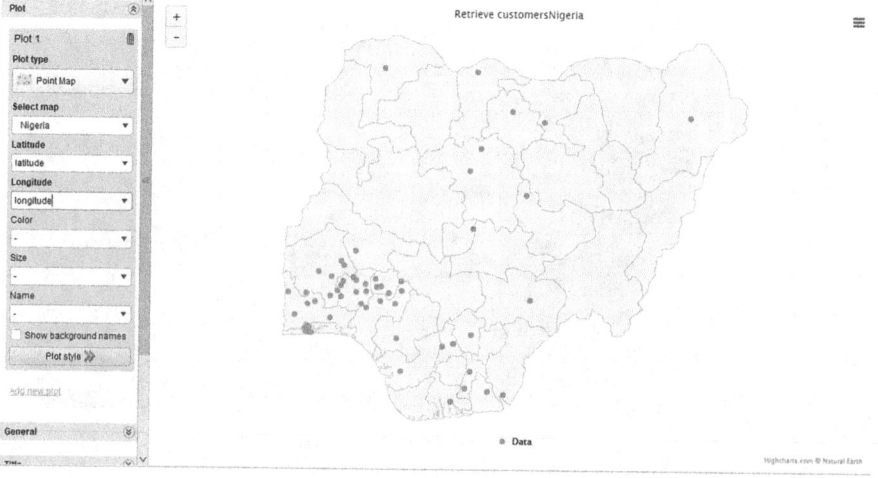

Figure 5-15. *Point map locating customers*

4. From Figure 5-15, we can easily make inferences about the location concentration of the customers in Nigeria.

5. Other types of maps can be explored such as the choropleth maps for displaying numerical values associated to regions (e.g., a country or a state) via a color gradient or the categorical maps used to visualize regions that belong to a number of distinct categories.[7]

It is important to say that for the purpose of demonstration, we have just selected some popular tools and used the most common setting for the visualization. There are other settings that can be used based on the goal of the analysis. Also, there are other more interactive visualization tools such as Tableau and Power BI which can be explored.

5.11 A Practical Business Problem I (Simple Descriptive Analytics)

This is the first out of the five business problems addressed in this book.

Problem scenario: In this problem, we use an ecommerce business as a case study. Let us assume that we have a business that sells clothes online. The data (SuperstoreNigeria.xls) used contains the record of sales. The task is to use descriptive analytics tools to describe what happened in the past and what things look like currently. Our intention is to be able to use the result to understand the business environment and make better sales decisions. The data contains the attributes: Row ID, Order ID, Order Date, Ship Date, Ship Mode, Customer ID, Segment, Country, State,

Postal Code, Region, Product ID, Category, Sub-Category, Sales, Quantity, Discount, Profit. In this problem, we intend answering the following questions:

1. What is the sum of sales for each subcategory for each region?

2. How does sales compare with profit from 2014 to 2017?

3. Which states are making high profits and which states are making a loss (use geographic map)?

4. What is the relationship between discount, sales, and profit?

5. In what category, subcategory, and State of Nigeria are the highest sales made?

Solution

1. *What is the sum of sales for each subcategory for each region?* The intention behind this question is to know what particular subcategory of their products gives them the highest sales and where (as regards region) this is happening. The first thing to do is to **Import** SuperstoreNigeria.xls and drag it to the design view, run the process, and in the *Results* interface, click visualization. To answer this question, we will use the bar chart. Figure 5-16a gives the setting for the answer to this question. From Figure 5-16b, which is the bar chart visualization, we can see that the East Region is having the highest total sales for categories: copiers,

machines, phones, storage, and book cases. The
West Region is having the highest total sales for
categories: supplies, accessories, tables, and chairs.
Overall, the West Region has the highest sales in
the chairs subcategory. This information can help
to know what products to keep producing in large
quantities in the future and which region to target.

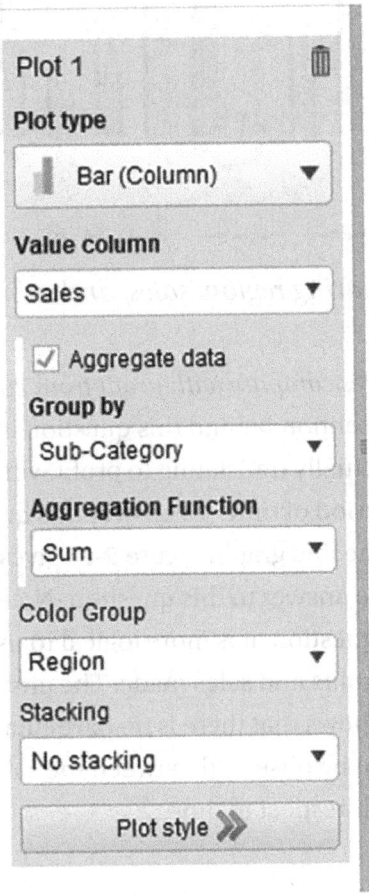

Figure 5-16a. *Settings for comparing region, sales, and subcategory*

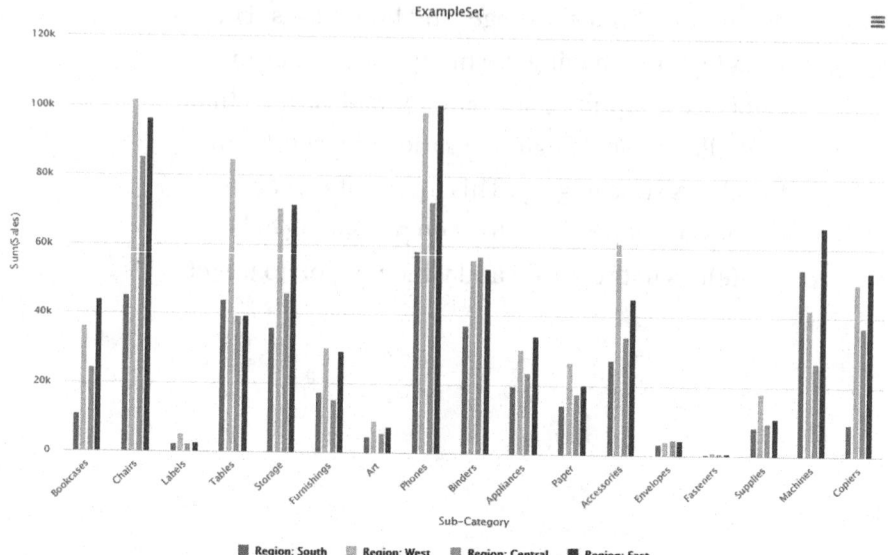

Figure 5-16b. *Comparing region, sales, and subcategory*

2. *How does sales compare with profit from 2014 to 2017?* The intention behind this question is to know if sales are actually translating to profit within a particular period of time. To answer this question, we will use the line graph. Figure 5-17 gives the setting for the answer to this question. Note that to answer this question, it is more logical to use the sum of the profits and sales made. The answer in Figure 5-17 shows that there is no particular linear trend that can be observed between the sales and the profit with respect to time.

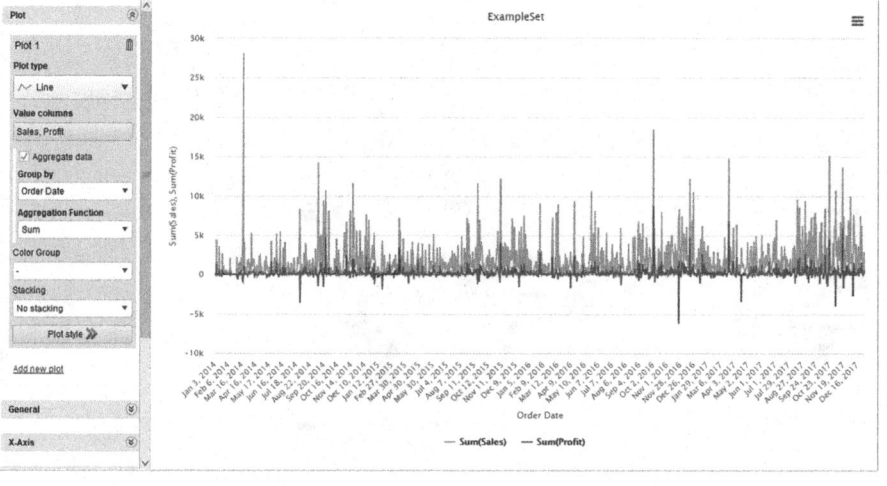

Figure 5-17. *Comparing profits and sales over time*

3. *Which states are making high profits and which*
    states are making a loss? The answer to this question
    can help to determine which state to focus on,
    for example, when marketing, so as to get more
    profit, and if, for example, they are making loss in
    a state and spending more marketing resources
    in that state, it might be wise to further investigate
    why this is so. To answer this question, we will use
    geographic maps. Figure 5-18 gives the setting for
    the answer to this question. From Figure 5-18, we
    can see that Lagos has the highest sales in total.

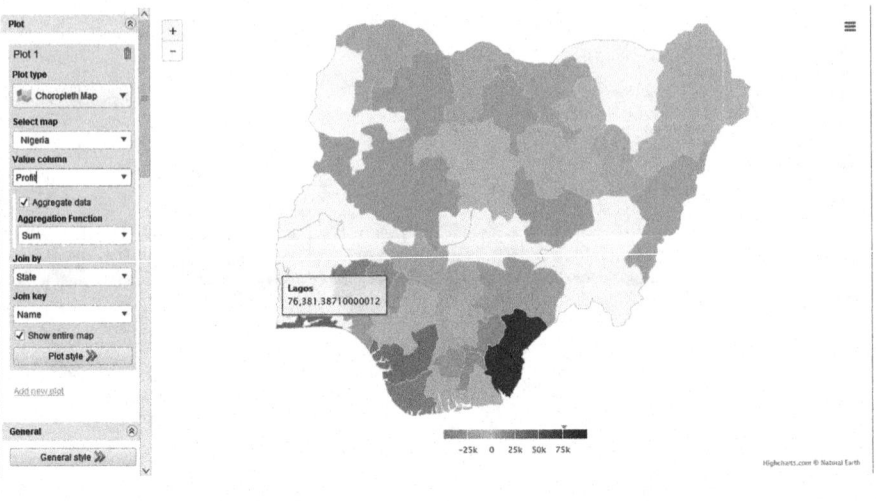

Figure 5-18. *Profit and loss per region*

4. *What is the relationship between the discount given and the profit made?* The intention behind this question is to know if the discount offered is translating to the profit that is being made, given that the whole idea behind discount is to make more sales and eventually make more profit for the organization. The answer to the question is revealed by using the settings in Figure 5-19a. To answer this question, we are using the total discount and total sales for each unique Row ID, to get the holistic view of the result.

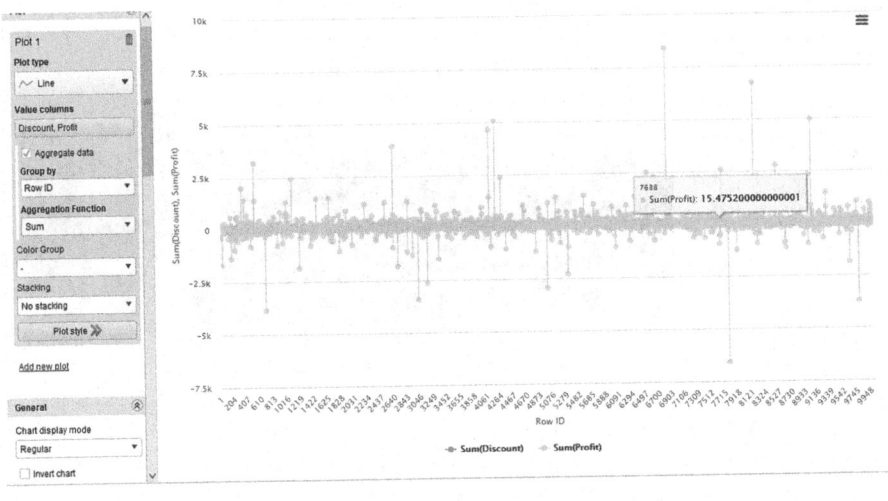

Figure 5-19a. *Discount and profit (using the original values)*

Unfortunately, due to the discount having smaller values than profit, it is almost not showing up in the graph. We have to select the logarithmic scale to see the discount as revealed in Figure 5-19b. From Figure 5-19b, we see that there is no linear trend that is observed between the discount and the profit.

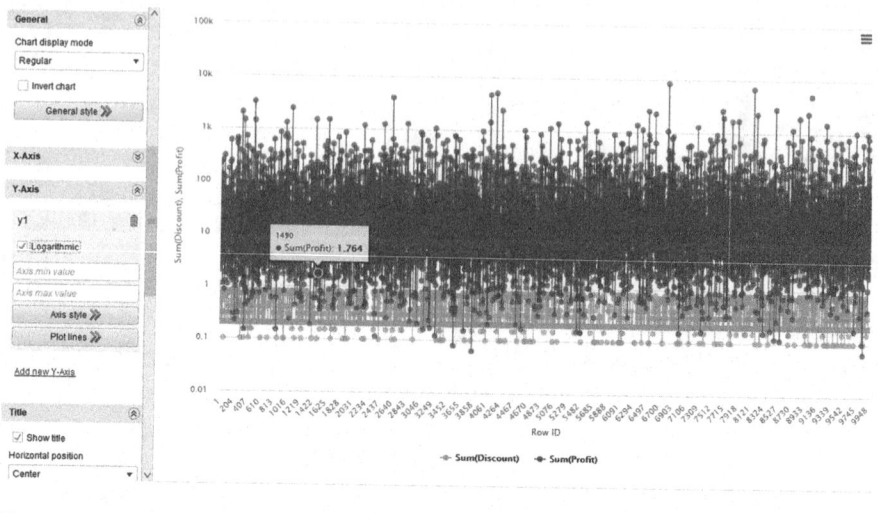

Figure 5-19b. *Discount and profit (using the normal values)*

5. *In what category or subcategory is the highest sales made?* The intention behind this question is to know the category, subcategory in particular, that is giving the highest sales. The answer to the question is revealed in the settings in Figure 5-20. The subcategory giving the highest sales is the phones, and it belongs to the category of technology. The answer to this question can help reveal the bestselling subcategory and category irrespective of the region, etc.

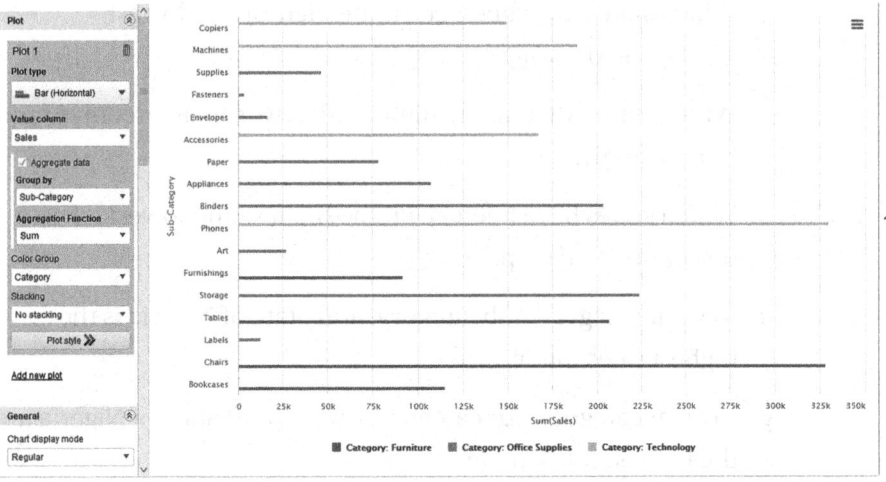

Figure 5-20. *Category, subcategory, and sales*

5.12 Problems

1. This exercise is the continuation of the practical business problem I (Section 5.11). The intention is to discover more insights, using the SuperstoreNigeria.xls to answer the following questions. (Note: Some questions might require more interactive visualization tools like Tableau.)

    a. What is the sum of sales for each subcategory and how does this relate with the profit, that is, which subcategory has the highest and lowest profit. Use color to indicate this.

    b. What is the sum of sales for each segment and how does this relate with the quantity, that is, which segment has the highest and lowest quantity produced. Use color to indicate this.

c. Which states have been given the highest and lowest discount? (Use map.)

d. Which states are making high profits (wordmap, treemap, and bubble map)?

e. Which states have been given the highest and lowest discount? (Use map.)

f. In what category, subcategory, and State of Nigeria is the highest sales made?

g. In what category, subcategory, region, and State of Nigeria are the highest sales made?

h. For the category named furniture, that the ship mode attribute is standard class, what is the total profit made and the total sales made. Make sure that your output is titled appropriately. Total Sales and Total Profit for Category and Ship Mode.

i. What is the total quantity sold and total sales made for office supplies using standard class ship mode.

j. Based on the category and subcategory, compare the sales in the East and West Regions.

2. You have been provided with a data named *chapter5Assign.xls*. It consists of the transactional data of a Nigerian noodle retail business (*data is fictitious, created just for the assignment*). Assuming this business intends on obtaining recommendation for an effective marketing campaign based on the result of descriptive analytics and you have been employed as the data scientist, present your result to the management in the form of the dashboard.

5.13 References

1. Graphical Data Analysis with R by Antony Unwin, published by CRC Press, 2015.

2. Jim Frost (March 2022) Using Histograms to Understand Your Data, https://statisticsbyjim. com/basics/histograms

3. Michael Galarnyk (September 2018) Understanding Boxplots, https://towardsdatascience.com/ understanding-boxplots-5e2df7bcbd51

CHAPTER 6

Predicting Numerical Outcomes

In this chapter, we will explore the popular techniques used for prediction, particularly in the retail business. The approach used in explaining these techniques is to use them in solving a business problem. The business problem to be addressed is the sales prediction problem which is common in the retail business. The chapter first explains the fundamental concept of prediction techniques; next, we look at how such techniques are evaluated. After this, we describe the business problem we intend to solve. We then pick each of the selected techniques one by one and explain the algorithms involved and how they can be used to solve the problem described. The prediction techniques used are the multiple linear regression, the regression trees, and the neural network. To conclude the chapter, we compare the results of the three algorithms and conclude on the problem in question. In this chapter, therefore, the analytics product offered is the sales prediction problem for small retail businesses.

6.1 Introduction

In Section 1.3 under the analytics journey, it was mentioned that there are three major types of analytics: descriptive, predictive, and prescriptive. In this chapter, the predictive analytics will be used to solve the sales

© Afolabi Ibukun Tolulope 2022
A. I. Tolulope, *Data Science and Analytics for SMEs*,
https://doi.org/10.1007/978-1-4842-8670-8_6

prediction problem for small retail businesses. To lay a good foundation for predictive analytics, we will explain some fundamental concepts.

Machine learning is used to investigate how computers learn or improve performance based on data. In machine learning, computer programs automatically learn to recognize complex patterns and make intelligent decisions based on data.[3]

Supervised learning: This is the process of feeding data to an algorithm, in which an output variable of interest is known, and the algorithm "learns" how to predict this value with fresh records when the output is unknown. Predictive analytics falls under the supervised. There are two types of predictive analytics, predicting categorical outcome (classification), for example, buy or no buy, or predicting numerical outcome (prediction), for example, stock price. Examples of predictive algorithms include multiple linear regression, regression tree (decision tree), logistic regression, SVM (support vector machine) for regression, and so on. Examples of classification methods include Naïve Bayes, Bayesian network, classification tree (decision tree), KNN, SVM (support vector machine), and so on. In this chapter, we will focus on the kind of predictive analytics that deals with predicting numerical outcome. The historical data used for predicting numerical outcome is labeled in the sense that all the rows in the data will end with a target attribute which is the attribute we intend predicting. In predicting either numerical or categorical outcome, the data needs to be portioned to 70% or 60% for training and 30% or 40% for testing, respectively. There are some algorithms like the KNN (K-nearest neighbor) that require a different kind of partitioning because there is a need to use a partition to determine the best k. This will be covered in Chapter 7.

In business, prediction can be used to achieve the following, for example:

- Predicting client credit card behavior based on demographics and previous activity patterns

- Using historical frequent flyer information to predict vacation travel spending

- Using historical data, product, and sales information to predict helpdesk personnel requirements

- Based on historical data, predicting sales through product cross-selling

- Predicting the effect of discounts on retail sales

Classification can be used to achieve the following, for example:

- To determine which loan applicants are dangerous or safe to lend to.

- In marketing, for example, classification can help to know if a customer will buy a product or not.

- Classification is very valuable in determining customer loyalty; in particular, we can predict if a customer will churn or not.

Unsupervised learning: This type of analysis entails attempting to discover patterns in the data rather than projecting the desired output value. Clustering falls under the unsupervised learning approach to machine learning, and there are many more. In this approach, the raw, unlabeled, and partitioned data is fed into the algorithm, and the output is labeled or classified data.

6.2 Evaluating Prediction Models

In this section, we look at how to evaluate prediction models, which is the main focus of this chapter. Several measures are used to evaluate prediction performance. These metrics are based on a validation or test dataset, which is more objective than the training dataset. The validation dataset isn't used to choose predictors or calculate model parameters. The prediction error (e) is the discrepancy between the outcome value and the predicted outcome value for the record i.[1,3]

$$e_i = y_i - \hat{y}_i \qquad \text{(Eq. 6.1)}$$

Some numerical measurements of prediction accuracy are as follows:

- *MAE (mean absolute error/deviation)*: This represents the average absolute error's magnitude.

$$\frac{1}{n}\sum_{i=1}^{n}|e_i| \qquad \text{(Eq. 6.2)}$$

- *Mean error*: This measure is similar to the MAE, except that it preserves the sign of the errors, allowing negative errors to cancel out positive errors of equal magnitude.

$$\frac{1}{n}\sum_{i=1}^{n}e_i \qquad \text{(Eq. 6.3)}$$

- *MPE (mean percentage error)*: This is a percentage score that indicates how far predictions differ from actual values (on average), taking into account the error's direction.

$$100\times\frac{1}{n}\sum_{i=1}^{n}\frac{e_i}{y_i} \qquad \text{(Eq. 6.4)}$$

- *MAPE (mean absolute percentage error)*: This metric gives a percentage score for how far predictions differ from actual values (on average).

$$100 \times \frac{1}{n} \sum_{i=1}^{n} \left| \frac{e_i}{y_i} \right| \qquad \text{(Eq. 6.5)}$$

- *RMSE (root mean squared error)*: This is comparable to the standard error of estimate in linear regression; however, it is calculated using validation data instead of training data. Its units are the same as the outcome variables.

$$\sqrt{\sum_{i-1}^{n} e_i^2} \qquad \text{(Eq. 6.6)}$$

A lift chart is a graphical representation of a model's prediction performance, especially when the purpose is to rank the predicted results. (For more explanations, check Galit et al. (2018), Chapter 5.)

6.3 Practical Business Problem II (Sales Prediction)

For this practical business problem, we will use a large Toyota car dealership company. The way the business runs is that if you want to buy a new Toyota car, you have an option to sell your old car to the company and add money to what you get and buy a new one. The company now aims to entice more consumers to come in and sell their used Toyota Corollas by running a promotion in which it will pay top dollar for used Toyota Corollas. The company's goal is to resell the used ones that were purchased and make a profit. So the business analyst's job is to predict the price at which the company will purchase these old cars (from their customers) in order to make a profit. Since this book looks at solving the business problems from the consulting approach, we try to present the solution to resemble the typical way you would approach it as a consultant. To do this, we explore three suitable and most popular algorithms (multiple linear regression, regression trees, and neural network) and compare the results to make our conclusions. It is important to say here that selecting the algorithms used here is not the only way that this experiment can be carried out, we can try other approaches, particularly in terms of the attribute combinations, algorithm settings, and algorithm types. The ultimate goal is to get the best performance while avoiding bias in data.

The data for this practical business problem is named *ToyotaCorollaData.xlsx*, and it offers information about past used cars purchased by the company. The data description is as follows:

Variable	Description
Id	*Unique identification*
Price	*Offer price*
Age	*Age in months*
KM	*Kilometers*
Fuel_Type	*Fuel type (Petrol, Diesel, CNG)*
HP	*Horsepower*
Met_Color	*Metallic color*
Automatic	*Automatic (Yes = 1, No = 0)*
CC	*Cylinder volume in cubic centimeters*
Doors	*Number of doors*
Quarterly_Tax	*Quarterly tax*
Weight	*Weight in kilograms*

Regardless of the prediction algorithm that will be used, the very first thing is to perform basic data preprocessing and explore the data using the necessary visualization tools for the purpose of prediction. The following preprocessing task will be carried out:

1. Statistics and missing value issues

2. Outlier detection

3. Visualization

4. Data transformation

5. Correlation analysis

All these steps have been handled one way or the other in numbers 1–10 under the practical demonstration.

Practical demonstration

1. Create a repository named **Sales Prediction**, create two folders in the repository (data and process), import the data named *ToyotaCorollaData.xlsx*, and save into the data folder in the repository.

2. When loading in ToyotaCorollaData.xlsx, on the **Format your columns** (data types) interface, select **Replace error with missing values**. Also on the same interface, you can edit the data types if necessary, for example, maybe a categorical data has been given an integer; you can change it to nominal or polynominal. Click **next** and **finish** the data import process. The final interface after loading the data should look like Figure 6-1.

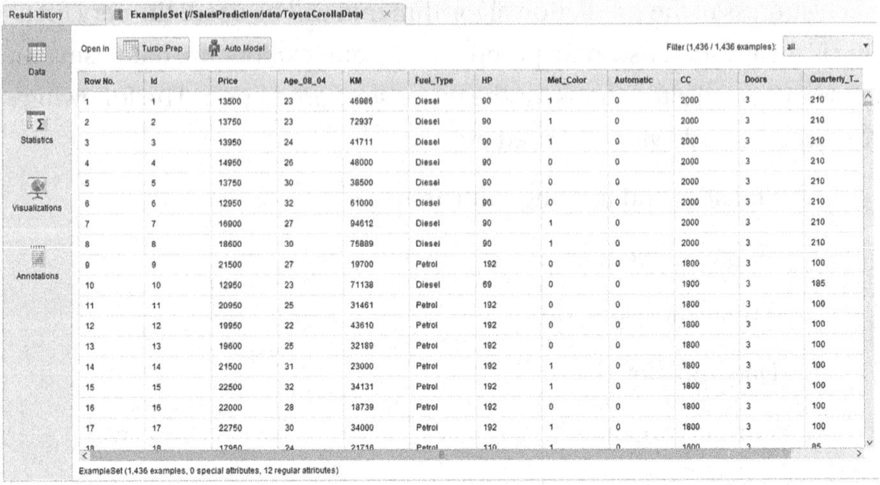

Figure 6-1. *Load data*

3. To view the statistics of your data and also check the missing values, click **Statistics** in Figure 6-1. Note that there are no missing values because the data we used has no missing values, but if there is, you will need to deal with it with the process explained in Section 2.4.

4. The **Statistics** interface is revealed in Figure 6-2. By clicking the attribute under the statistics, you will see the default statistical visualization of each attribute (histogram). At this point, you want to decide what to do about each of the attributes based on the analytics goals (refer to Section 2.5). In our current example, since the goal is prediction, we will perform our visualization using the prediction visualization guides in Section 2.5.

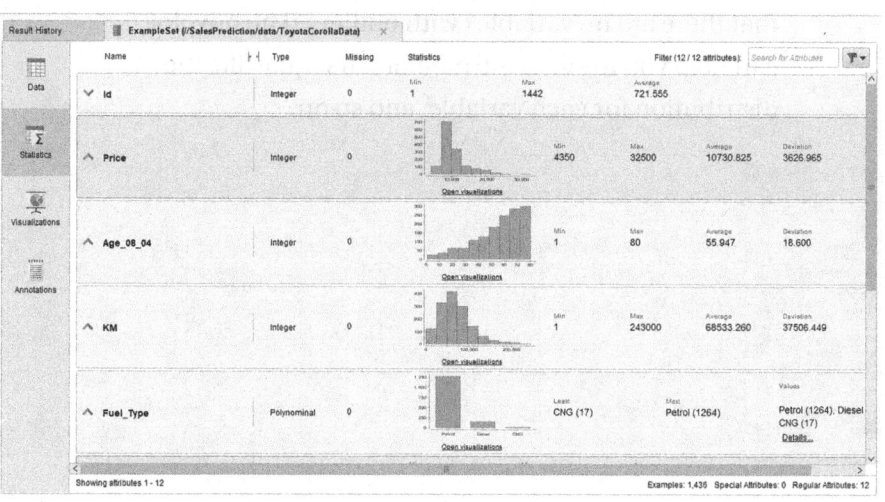

Figure 6-2. *Summary statistics*

5. Use a histogram/boxplot to determine any needed transformation (plot the outcome variable against the numerical predictors). Checking the histogram in Figure 6-2 for all the variables, we will notice that some numerical variables are skewed to the right or left, particularly the price, age, and weight. The next thing will be to do some form of data transformation on the data, for example, applying logarithmic functions. For the purpose of this example, we will not do this as we will still be using the necessary attribute reduction for each of the algorithms selected. Note that you could actually create the histograms for yourself using the ***Visualizations*** tab. Figure 6-3 shows the boxplot of some of the numerical variables which we are using to find out if the data has outliers and so on. If you do this for all the other numerical attributes, you will find out

that there are no variables with outliers. The boxplot can help you to study the median, first quartile, the distribution for each variable, and so on.

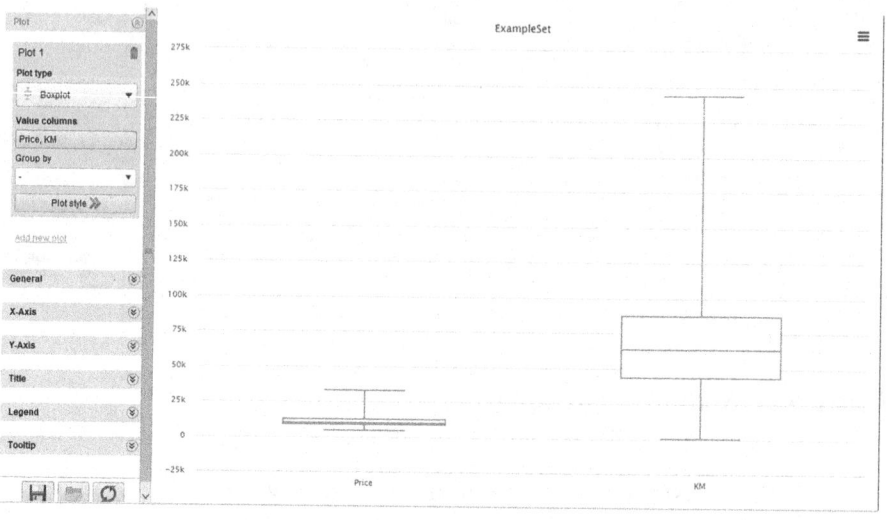

Figure 6-3. *Histograms*

6. Use a scatter plot to study the relationship between the numerical predictors. When you do this, you are looking at relationships between them that could affect your analysis negatively or positively. Figure 6-4 is the scatter plot of attribute KM and Price; this reveals a slight relationship between the two attributes. As the price is reducing, the KM is increasing. This relationship is not set to affect our experiment negatively, so we can leave the two variables, but if, for example, we discover that the two variables are highly correlated, we will have to remove one as this could affect the multiple linear regression algorithm, for example. This combination

should be tried for all the numerical variables.
Note the Met_Color and Automatic are categorical
variables and should be converted when loading
the data.

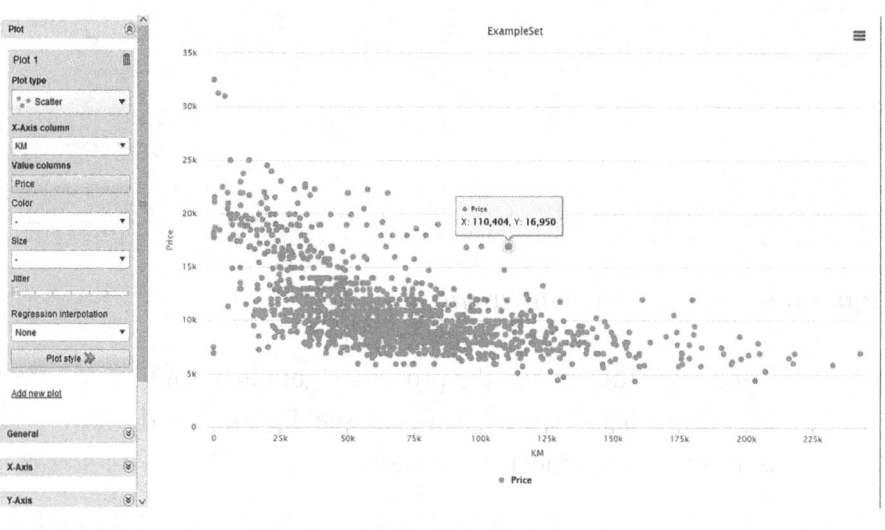

Figure 6-4. *Scatter plots*

7. You can try other guides for visualizing for
   prediction in Section 2.5 and make your
   conclusions. We will use only the preceding two
   guides in this practical demonstration.

8. For the purpose of our prediction task, we need
   to convert the categorical variable ***Fuel_Type*** to
   dummies. To do this, in the design view, create
   the process as revealed in Figure 6-5. The operator
   Nominal to Numerical can be searched for in the
   Operators window.

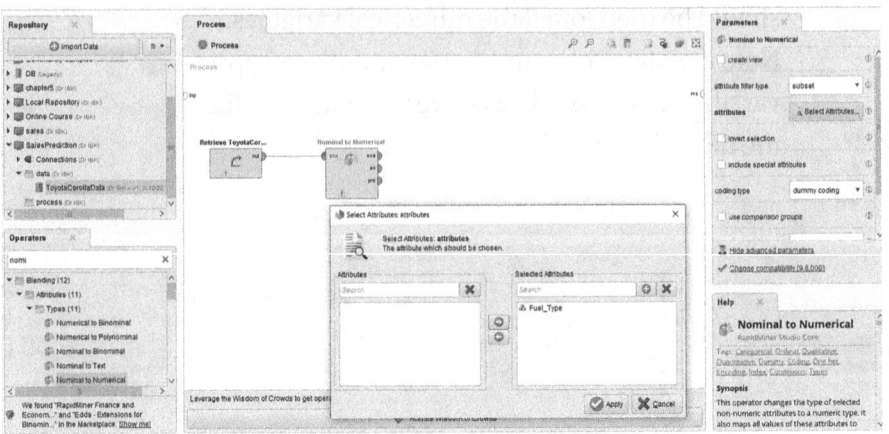

Figure 6-5. *Nominal to Numerical*

Connect to the ***res***; run the process to get the output
in Figure 6-6. In Figure 6-6, we can see the converted
dummies for the fuel type category.

Row No.	Fuel_Type = ...	Fuel_Type = ...	Fuel_Type = ...	Id	Price	Age_0
1	1	0	0	1	13500	23
2	1	0	0	2	13750	23
3	1	0	0	3	13950	24
4	1	0	0	4	14950	26
5	1	0	0	5	13750	30
6	1	0	0	6	12950	32
7	1	0	0	7	16900	27
8	1	0	0	8	18600	30
9	0	1	0	9	21500	27
10	1	0	0	10	12950	23
11	0	1	0	11	20950	25
12	0	1	0	12	19950	22
13	0	1	0	13	19600	25

Figure 6-6. *Output Nominal to Numerical*

9. To get the correlation matrix for the data, use the settings in Figure 6-7. The ***Correlation Matrix*** can be searched for in ***Operators***. Run the process to get the correlation matrix in Figure 6-8. Note that to know which method was used for the correlation matrix, you need to read the ***help*** of the operator. The result is shown in Figure 6-8.

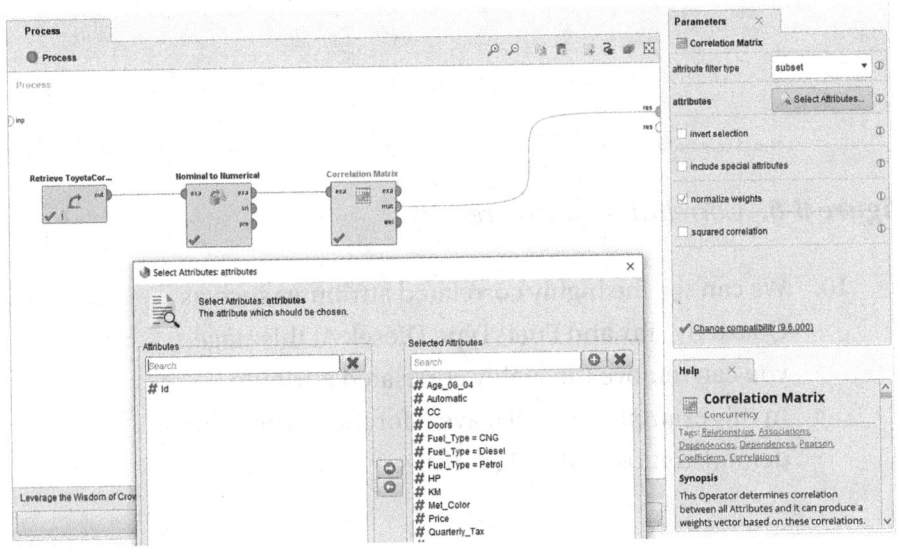

Figure 6-7. *Correlation matrix process*

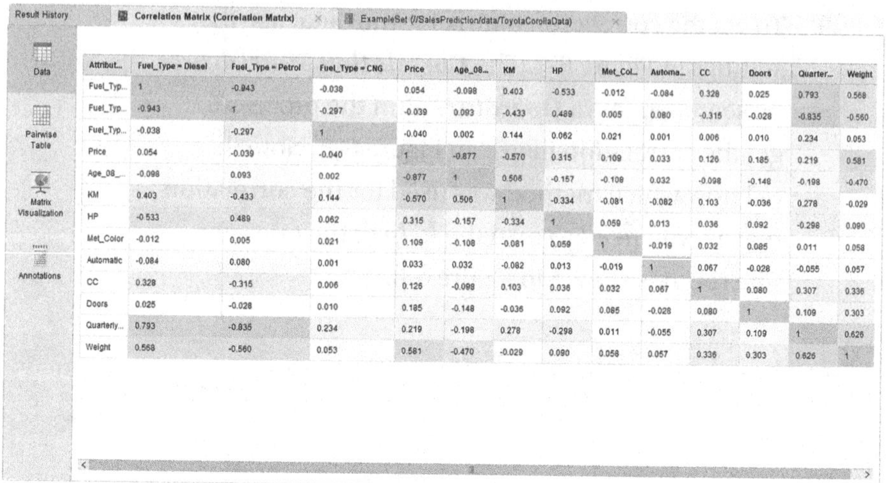

Attribut...	Fuel_Type = Diesel	Fuel_Type = Petrol	Fuel_Type = CNG	Price	Age_08...	KM	HP	Met_Col...	Automa...	CC	Doors	Quarter...	Weight
Fuel_Typ...	1	-0.843	-0.038	0.054	-0.098	0.403	-0.533	-0.012	-0.084	0.328	0.025	0.793	0.568
Fuel_Typ...	-0.943	1	-0.297	-0.039	0.093	-0.433	0.489	0.005	0.080	-0.315	-0.028	-0.835	-0.560
Fuel_Typ...	-0.038	-0.297	1	-0.040	0.002	0.144	0.062	0.021	0.001	0.006	0.010	0.234	0.053
Price	0.054	-0.039	-0.040	1	-0.877	-0.570	0.315	0.109	0.033	0.126	0.185	0.219	0.581
Age_08_...	-0.098	0.093	0.002	-0.877	1	0.506	-0.157	-0.108	0.032	-0.098	-0.148	-0.198	-0.470
KM	0.403	-0.433	0.144	-0.570	0.506	1	-0.334	-0.081	-0.082	0.103	-0.036	0.278	-0.029
HP	-0.533	0.489	0.062	0.315	-0.157	-0.334	1	0.059	0.013	0.036	0.092	-0.298	0.090
Met_Color	-0.012	0.005	0.021	0.109	-0.108	-0.081	0.059	1	-0.019	0.032	0.085	0.011	0.058
Automatic	-0.084	0.080	0.001	0.033	0.032	-0.082	0.013	-0.019	1	0.067	-0.028	-0.055	0.057
CC	0.328	-0.315	0.006	0.126	-0.098	0.103	0.036	0.032	0.067	1	0.080	0.307	0.336
Doors	0.025	-0.028	0.010	0.185	-0.148	-0.036	0.092	0.085	-0.028	0.080	1	0.109	0.303
Quarterly...	0.793	-0.835	0.234	0.219	-0.198	0.278	-0.298	0.011	-0.055	0.307	0.109	1	0.626
Weight	0.568	-0.560	0.053	0.581	-0.470	-0.029	0.090	0.058	0.057	0.336	0.303	0.626	1

Figure 6-8. *Correlation matrix results*

10. We can see the highly correlated attributes such as Quarterly_Tax and Fuel_Type Diesel. At this stage, you can remove the highly correlated attribute. In this example, we will leave it for the purpose of demonstration of the algorithms.

6.4 Multiple Linear Regression

The multiple linear regression (MLR) model illustrates the link between the dependent variable and several independent variables (two or more). The variance (i.e., R^2), as well as the unique contribution of each independent variable, are expressed in the MLR model's output. The regression line is the name given to the regression equation. The coefficient of determination, often known as R^2, provides the proportion of the variability of the independent variables (X) and quantifies the ability to predict an individual Y using these independent variables.

Equation 6.7 represents the regression equation. x1, x2,...xp are independent variables in this equation, while 0...p are coefficients and ϵ is the noise or unexplained element. Y stands for the dependent variable, which is the numerical attribute we want to predict. Data is utilized to estimate the coefficients, and data is also used to quantify the noise.

$$Y = \beta_0 + \beta_1 X_1 + \beta_2 X_2 + \beta_3 X_3 + \ldots\ldots\ldots + \beta_p X_p + \epsilon \qquad (Eq.\ 6.7)$$

The data is also utilized to evaluate model performance if the task at hand is to predict a numerical outcome. If the graph is skewed to the right or left when visualizing with the histogram, we may need to convert the independent variables using data transformation techniques (e.g., logarithmic form [log(X)). Note that the dependent variable does not have to be normally distributed, but the model's residuals have to be normally distributed for the MLR model to be valid. Furthermore, because MLR is sensitive to multicollinearity, there should be no multicollinearity, which means that the independent variables should not be highly correlated. The adjusted R^2, which is defined in Equation 6.8, is a popular method for evaluating MLR models.

$$R^2_{adj} = 1 - \frac{n-1}{n-p-1}\left(1 - R^2\right) \qquad (Eq.\ 6.8)$$

- p is the number of variables to be predicted.

- R^2 denotes the model's fraction of explained variability (in a model with a single predictor, this is the squared correlation).

- Higher R^2_{adj} values, like R^2, suggest greater fit.

Unlike R^2, which ignores the number of predictors used, R^2_{adj} penalizes the number of predictors used. For a more detailed discussion of the multiple linear regression model, see Galit et al. (2018).[3] The validation/test dataset is used to determine MLR's prediction performance. These are dataset partitions that are not meant to be used to pick predictors or calculate model parameters.

Variable selection in linear regression

Choosing the right form to use is determined by domain knowledge, data availability, and the required predictive capability. It's important to be cautious before employing all of the available variables to build your model because collecting a full complement of predictors for future forecasts can be costly or impossible. Furthermore, we may be able to measure fewer predictors more reliably (e.g., in surveys), and the more predictors we have, the more likely we are to have missing values in the data. To limit the number of predictors, there are a few options. To reduce the number of predictors, one technique is to employ domain knowledge. The utilization of computational power and statistical performance measures is another option. Some of these approaches are as follows:

- *Exhaustive search*: Evaluate all subsets of predictors and each subset model, using certain criteria to evaluate the models.

- *Popular subset selection algorithms*: This process uses an iterative partial search through the space of all potential regression models. Forward selection, backward elimination, and stepwise regression are examples of these.

- Principal component analysis, for example, is particularly useful when subsets of observations are measured on the same scale and are highly associated. If the number of variables is enormous, principal component analysis can help to identify a few variables that are weighted linear combinations of the original variables and maintain the majority of the original set's information. It's most commonly used with numerical variables. Check out Galit et al. (2018) for further information.

Sales prediction problem – MLR

1. The first thing we are going to do is to create a new process in RapidMiner and load the data again; we call it *ToyotaCorollaData2* (we will exclude the ID column). This can be seen in Figure 6-9.

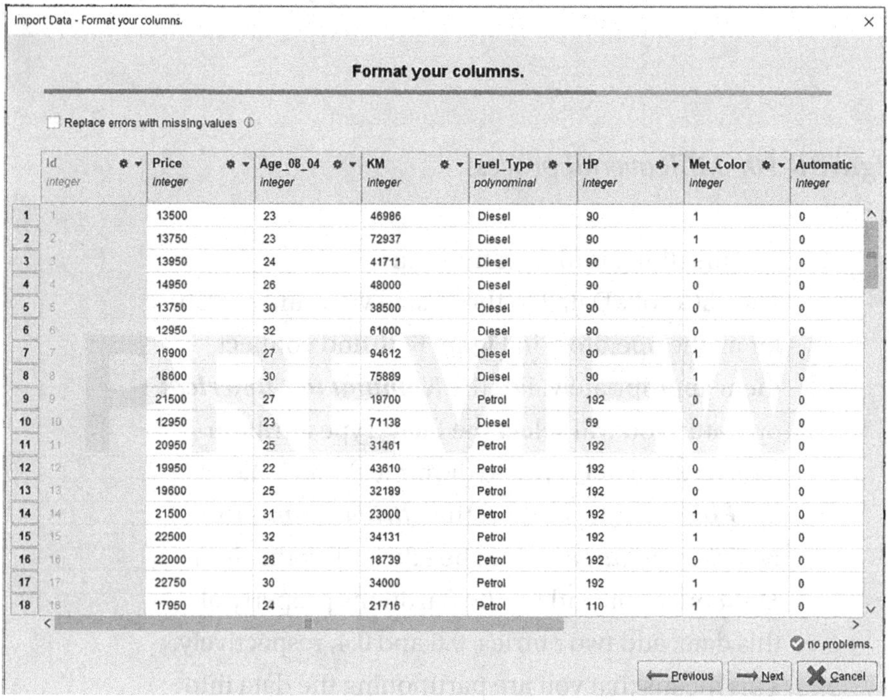

Import Data - Format your columns.   ×

Format your columns.

☐ Replace errors with missing values ⓘ

Id *integer*	Price *integer*	Age_08_04 *integer*	KM *integer*	Fuel_Type *polynominal*	HP *integer*	Met_Color *integer*	Automatic *integer*
1	13500	23	46986	Diesel	90	1	0
2	13750	23	72937	Diesel	90	1	0
3	13950	24	41711	Diesel	90	1	0
4	14950	26	48000	Diesel	90	0	0
5	13750	30	38500	Diesel	90	0	0
6	12950	32	61000	Diesel	90	0	0
7	16900	27	94612	Diesel	90	1	0
8	18600	30	75889	Diesel	90	1	0
9	21500	27	19700	Petrol	192	0	0
10	12950	23	71138	Diesel	69	0	0
11	20950	25	31461	Petrol	192	0	0
12	19950	22	43610	Petrol	192	0	0
13	19600	25	32189	Petrol	192	0	0
14	21500	31	23000	Petrol	192	1	0
15	22500	32	34131	Petrol	192	1	0
16	22000	28	18739	Petrol	192	0	0
17	22750	30	34000	Petrol	192	1	0
18	17950	24	21716	Petrol	110	1	0

✅ no problems.

⬅ Previous   ➡ Next   ✖ Cancel

Figure 6-9. *Loading the data*

2. The overall process for running the multiple linear regression is given in Figure 6-10.

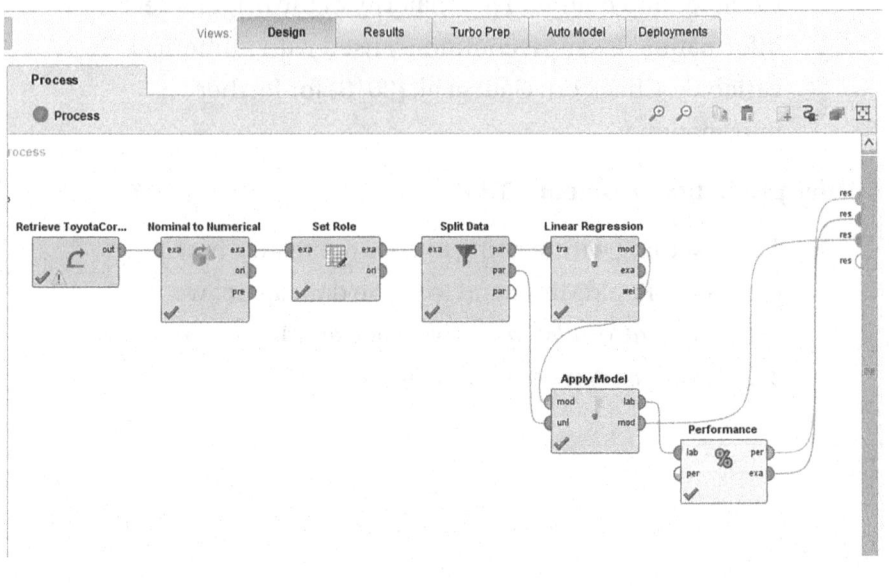

Figure 6-10. *MLR overall process*

3. The first thing is to drag the data ToyotaCorollaData2 to the design view and search for all the operators in Figure 6-10 and connect them appropriately. For the ***Nominal to Numerical*** operator, you will select the Fuel_Type attribute (converting Fuel_Type to dummies). Click the ***Set Role*** operator and set the *attribute name* to be *Price* and *target role* to be label. Click the ***Split Data*** operator, and for the partitions property of this data, add two entries, 0.6 and 0.4, respectively. This means that you are partitioning the data into 60% for the training and 40% for the validation. Click the ***Linear Regression*** operator; the feature

selection parameter though optional is M5 prime by default. This is an expert parameter. It indicates the feature selection method to be used during regression. The following options are available: none, M5 prime, greedy, T-Test, and iterative T-Test (more information in Rapidminer.doc). The *Apply Model* operator needs no parameter adjustment; it is used to apply the model obtained from the *Linear Regression* operator on the validation dataset. Note that the training dataset is what the linear regression model uses to build the model. For the parameters of the *Performance (Regression)* operator, simply leave as default, which is the *root mean square error*.

4. With the settings in Figure 6-10, we are expecting three outputs (i.e., three links to *res*). The first is the MLR model (Figure 6-11), the second is the model performance result (Figure 6-12), and the third is the data table revealing the predicted results (Figure 6-13) for the validation dataset.

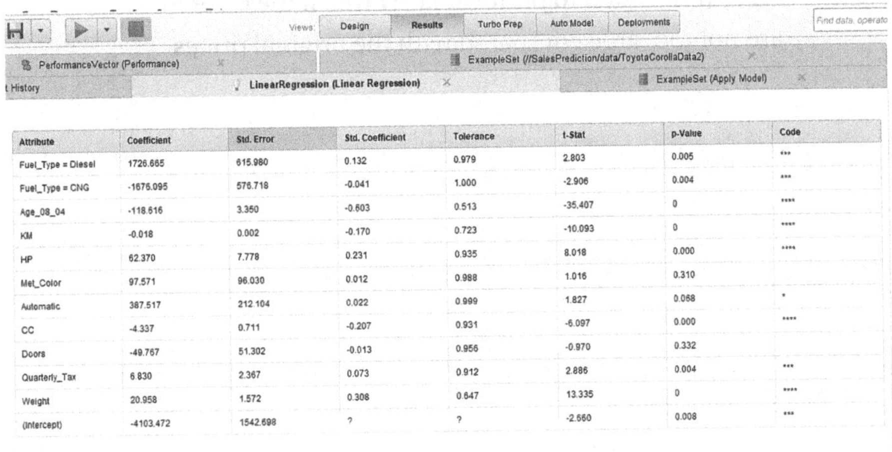

Attribute	Coefficient	Std. Error	Std. Coefficient	Tolerance	t-Stat	p-Value	Code
Fuel_Type = Diesel	1726.665	615.980	0.132	0.979	2.803	0.005	***
Fuel_Type = CNG	-1676.095	576.718	-0.041	1.000	-2.906	0.004	***
Age_08_04	-118.616	3.350	-0.603	0.513	-35.407	0	****
KM	-0.018	0.002	-0.170	0.723	-10.093	0	****
HP	62.370	7.778	0.231	0.935	8.018	0.000	****
Met_Color	97.571	96.030	0.012	0.988	1.016	0.310	
Automatic	387.517	212.104	0.022	0.999	1.827	0.068	*
CC	-4.337	0.711	-0.207	0.931	-6.097	0.000	****
Doors	-49.767	51.302	-0.013	0.956	-0.970	0.332	
Quarterly_Tax	6.830	2.367	0.073	0.912	2.886	0.004	***
Weight	20.958	1.572	0.308	0.647	13.335	0	****
(Intercept)	-4103.472	1542.698	?	?	-2.660	0.008	***

Figure 6-11. MLR model

- Note that this can be used to write the regression equation, and the third dummy for the *Fuel_Type* attribute is no more there. This is because when we convert to dummy, one of the dummies created is not used for modeling for some algorithms. Including the redundant dummy would cause the regression to fail, since the redundant dummy will be a perfect linear combination of the other two.

Figure 6-12. *MLR evaluation*

To conclude on the prediction performance, the algorithm is said to perform fairly well if the errors are not too large compared with the original prices.

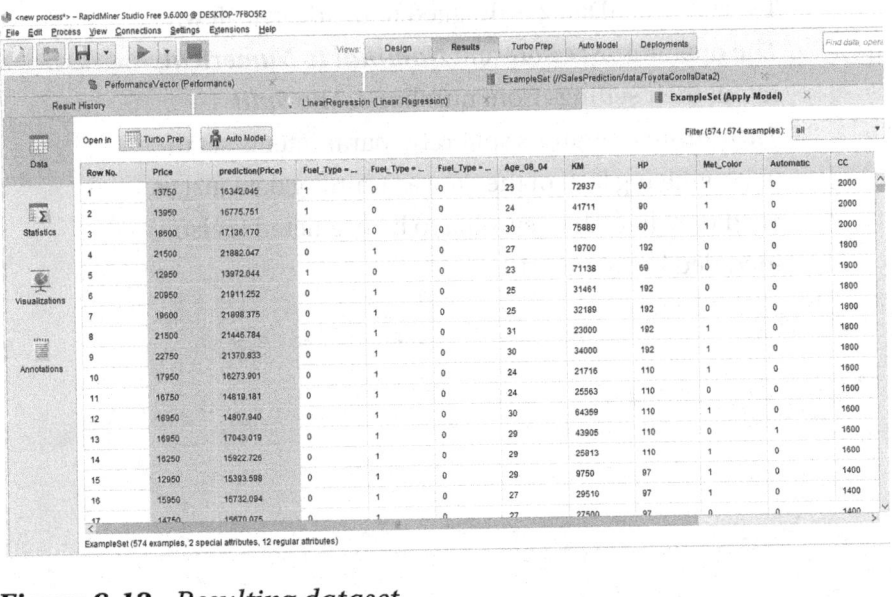

Figure 6-13. *Resulting dataset*

Figure 6-13 is the output of performing evaluation using the 40% validation dataset and comparing the predicted result and the original price in validation dataset.

5. Once we have established an optimal model for prediction, we can now use the model to predict for new data. Prepare the new dataset such that it does not have the outcome/target variable (ToyotaCorollaNewData.xlsx).

6. The first thing is to load the data (ToyotaCorollaNewData) into the data folder of the **SalesPrediction** repository. Remember to exclude the ID column of the ToyotaCorollaNewData dataset when importing the data. To predict for new records, we will use the settings in Figures 6-13a and 6-13b. Figure 6-13a is the main process. In this process, the

133

ToyotaCorollaData2 is dragged to the design view. The operators **Set Role** and **Nominal to Numerical** retain their settings from number 3. The **Split Validation** operator's split ratio parameter is set to 0.6, signifying 60% of the data as training data just as we did earlier. The new data to be predicted is also included in Figure 6-13a.

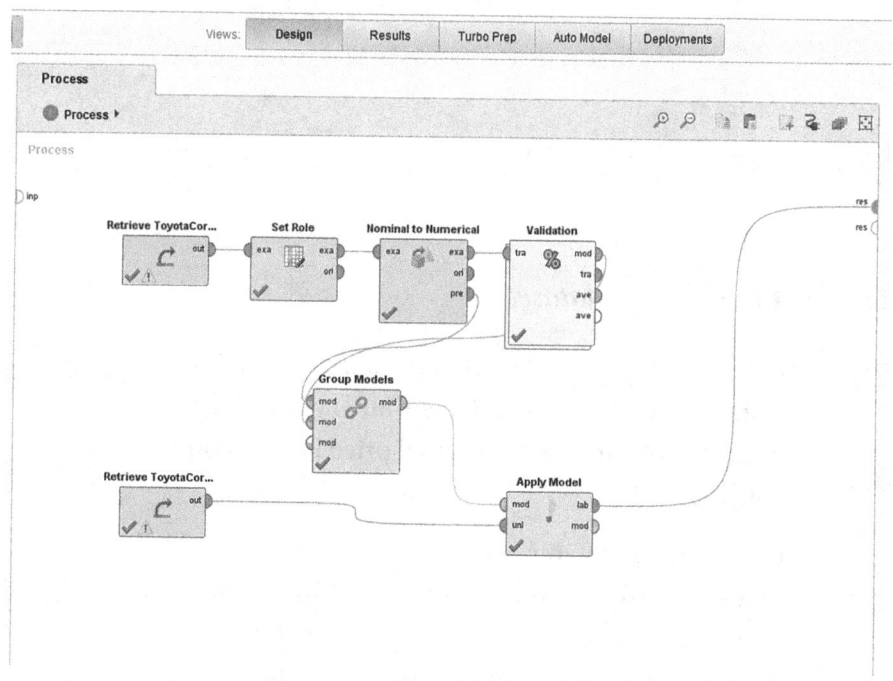

Figure 6-13a. *Predicting new data outer process*

7.   Figure 6-13b is the inner process which you will create by double-clicking the **Split Validation** operator. The **Linear Regression, Apply Model,** and **Performance (Regression)** operators are left as default settings just as earlier.

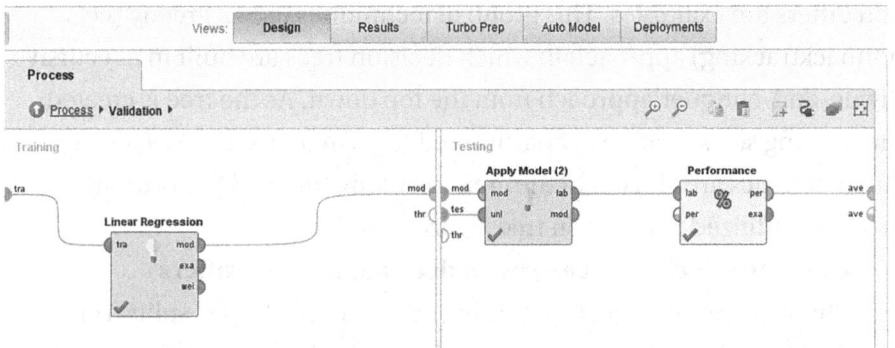

Figure 6-13b. *Inner process predicting new data*

8. The predicted prices for the new data loaded are given in Figure 6-14.

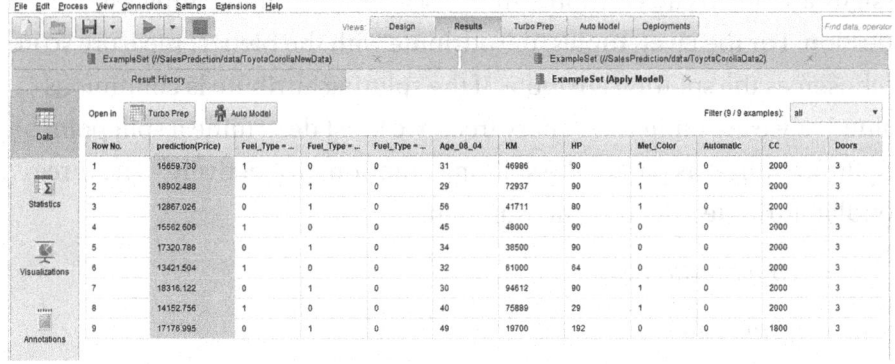

Figure 6-14. *Resulting dataset*

Note that we can also predict for the new record manually by using the regression equation obtained from the model in Figure 6-11.

6.5 Regression Trees

The family of decision tree classifiers includes regression trees. The data-driven method of decision tree classifiers necessitates a significant amount of data. ID3, C4.5, CART (Classification and Regression Trees), and others are examples. This group of techniques uses a greedy (i.e., nonbacktracking) approach in which decision trees are built in a recursive divide-and-conquer approach from the top down. As the tree is created, the training set is recursively partitioned into smaller subsets. The way characteristics are chosen in the creation of the tree and the pruning processes utilized in decision tree algorithms differ.

Attribute selection measures in decision tree classifiers

A heuristic for picking the splitting criterion that "best" splits a given data partition, D, of class-labeled training tuples into individual classes is an attribute selection measure. If we split D into smaller partitions, the optimum splitting requirement is for each partition to be pure. Because they determine how tuples at a given node are split, attribute selection measures are also known as splitting rules. The attribute selection measure assigns a score to each attribute that describes the training tuples in question. For the given tuples, the attribute with the highest measure score is chosen as the splitting attribute. If the splitting attribute is continuous-valued or we are limited to binary trees, we must determine a split point or a splitting subset as part of the splitting criterion, respectively.[2] The three popular attribute selection measures are

- *Information gain*

- *Gain ratio*

- *Gini index*

An example of a decision tree structure is given in Figure 6-15.

- A root node, branches, and leaf nodes are all included.

- Each internal node represents a test on an attribute.

- Each branch represents the test's result, and each leaf node represents a class.

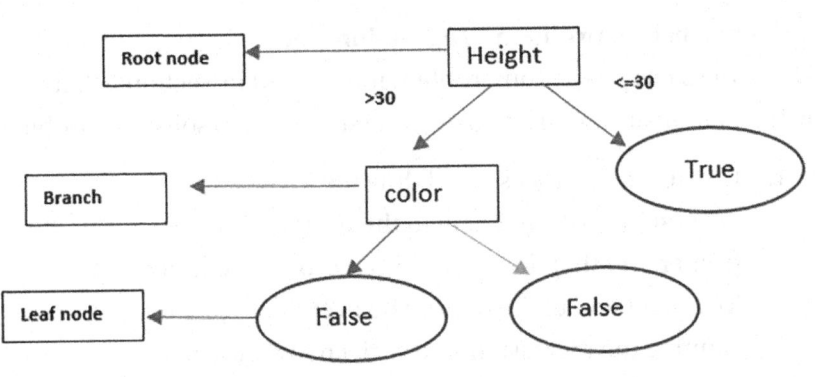

Figure 6-15. *Decision tree example*

Some of the benefits of employing decision tree classifiers include the fact that they do not require domain knowledge, are simple to understand, and are white boxes, meaning that we can quickly interpret decision tree models. This can be accomplished by extracting the decision tree's rules. Outliers aren't a problem for decision tree algorithms, and they can tolerate missing values.

Overfitting is a risk of building depth trees on the training data when utilizing decision trees. When final splits are based on a tiny number of records, this is said to happen. Overfitting will result in poor performance while dealing with new data. To avoid overfitting, we can use the tree depth parameter, the least number of records in a terminal node, and the minimal decrease in impurity to stop the tree from growing before it starts overfitting the data. In our data partitioning, we can additionally prune the tree or utilize cross-validation.

Regression trees

Both numerical and categorical outcomes can be predicted using decision tree algorithms. Regression trees are used when the expected outcome is a numerical value. When using a regression tree to make a prediction, the predictor data is utilized to "drop" the record down the tree until it reaches a terminal node. The average outcome value of the training records in that terminal node determines the value of the terminal node in regression trees.

Sales prediction problem – regression trees

Using the sales prediction problem introduced in Section 6.2, the following demonstrates how to use regression trees to solve the problem:

1. Create a new process, and drag the *ToyotaCorollaData2* data to the ***design view*** (remember that the ID column has been excluded in this version of the data). The overall process for running the regression tree is given in Figure 6-16.

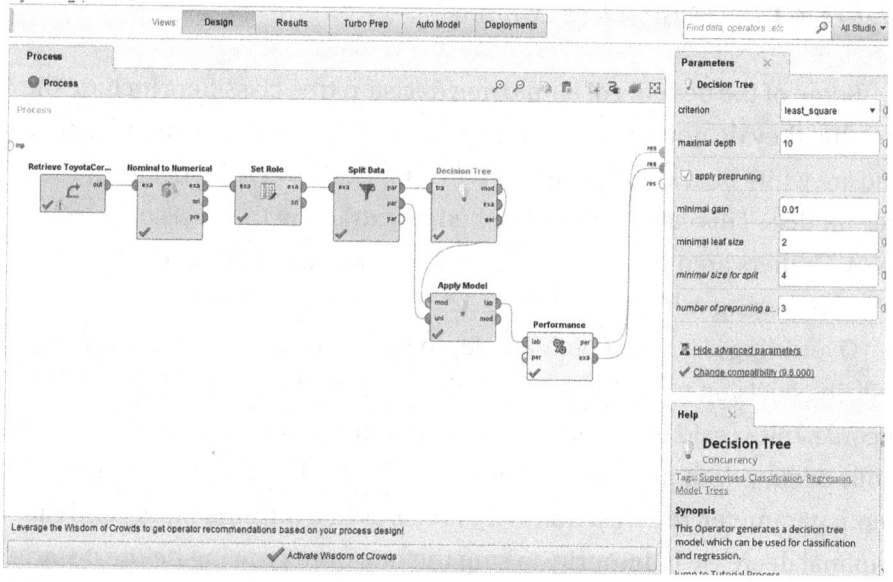

Figure 6-16. *Regression tree process*

2. For the ***Nominal to Numerical*** operator, you will select the Fuel_Type attribute (converting Fuel_Type to dummies). Click the ***Set Role*** operator and set the *attribute name* to be *Price* and *target role* to be label. Click the ***Split Data*** operator, and for the partitions property of this data, add two entries, 0.6 and 0.4,

respectively. Click the **Decision Tree** operator and
set the criterion parameter to least_square. Note
that least_square is the only criterion that allows
for numerical prediction. All other parameters
can be set based on their explanations in the help.
The **Apply Model** operator needs no parameter
adjustment; it is used to apply the model obtained
from the **Decision Tree** operator on the validation
dataset. For the parameters of the **Performance
(Regression)** operator, simply leave as default,
which is the *root mean square error.*

3. With the settings in Figure 6-16, we are expecting
    three outputs (i.e., three links to ***res***). The first is the
    regression tree model (Figure 6-16a), the second is
    the model performance result (Figure 6-16b), and
    the third is the data table revealing the predicted
    results (Figure 6-16c) for the validation dataset.

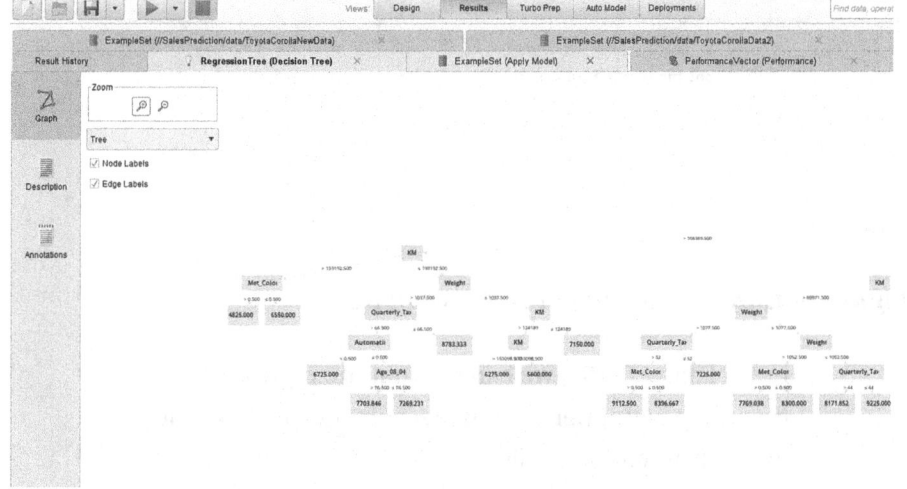

Figure 6-16a. *Some sections of the regression tree model*

Among the data-driven methods, trees are the most transparent and easy to interpret. Trees are based on separating records into subgroups by creating splits on predictors. These splits create logical rules that are transparent and easily understandable. Examples of such from the model in Figure 6-16a include

"IF Age_08_04 >=32.500

AND Age_08_04 > 56.500

AND Age_08_04 > 68.500

AND KM> 108389.500

AND KM> 193192.500

AND MET_Color>0.500

THEN Price is 4825.000"

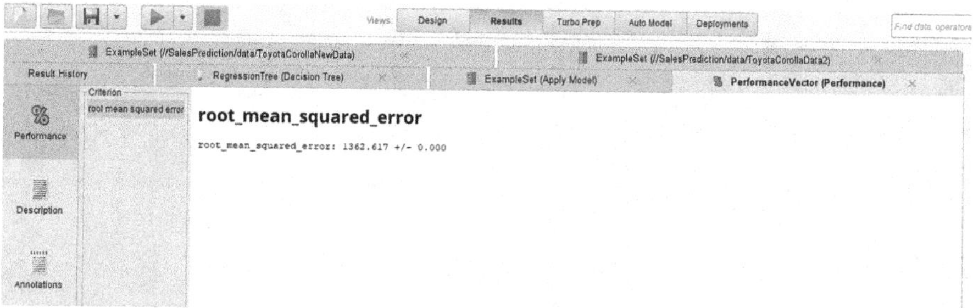

Figure 6-16b. *Evaluating the regression tree*

From Figure 6-16b, we can see that the regression tree gives better performance result compared to the multiple linear regression.

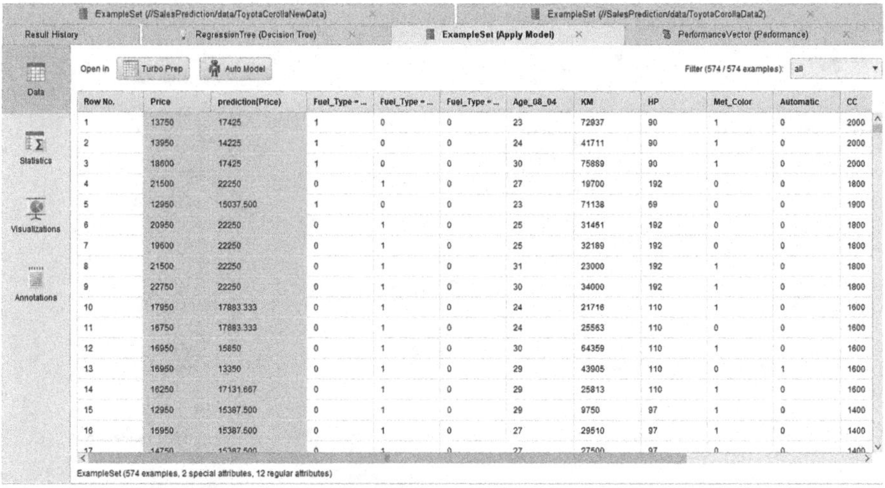

Row No.	Price	prediction(Price)	Fuel_Type - ...	Fuel_Type - ...	Fuel_Type - ...	Age_08_04	KM	HP	Met_Color	Automatic	CC
1	13750	17425	1	0	0	23	72937	90	1	0	2000
2	13950	14225	1	0	0	24	41711	90	1	0	2000
3	18600	17425	1	0	0	30	75859	90	1	0	2000
4	21500	22250	0	1	0	27	19700	192	0	0	1800
5	12950	15037.500	1	0	0	23	71138	69	0	0	1900
6	20950	22250	0	1	0	25	31461	192	0	0	1800
7	19600	22250	0	1	0	25	32189	192	0	0	1800
8	21500	22250	0	1	0	31	23000	192	1	0	1800
9	22750	22250	0	1	0	30	34000	192	1	0	1800
10	17950	17883.333	0	1	0	24	21716	110	1	0	1600
11	16750	17883.333	0	1	0	24	25563	110	0	0	1600
12	16950	15850	0	1	0	30	64359	110	1	0	1600
13	16950	13350	0	1	0	29	43905	110	0	1	1600
14	16250	17131.667	0	1	0	29	25813	110	1	0	1600
15	12950	15387.500	0	1	0	29	9750	97	1	0	1400
16	15950	15387.500	0	1	0	27	29510	97	1	0	1400
17	14750	15387.500	0	1	0	27	27500	97	0	0	1400

ExampleSet (574 examples, 2 special attributes, 12 regular attributes)

Figure 6-16c. *Resulting dataset*

Figure 6-16c is the output of evaluating on the 40% validation dataset and comparing the predicted result and the original price in the validation dataset.

4. For the regression tree, we will use the settings in Figures 6-17a (main process) and 6-17b (inner process) to predict for new records.

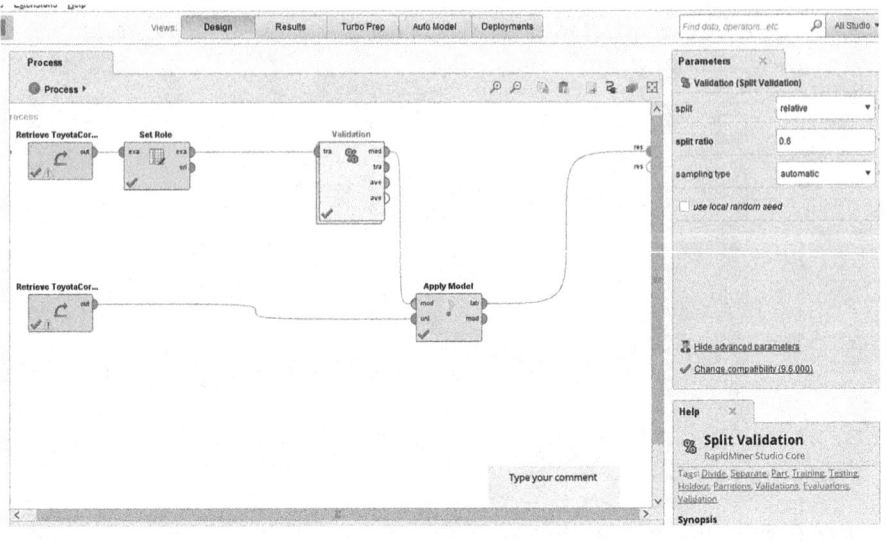

Figure 6-17a. *Regression tree main process*

5. Figure 6-17b is the inner process which you will
    create by double-clicking the ***Split Validation***
    operator. The ***Decision Tree***, ***Regression***, ***Apply
    Model***, and ***Performance (Regression)*** operators
    are left as default settings just as earlier.

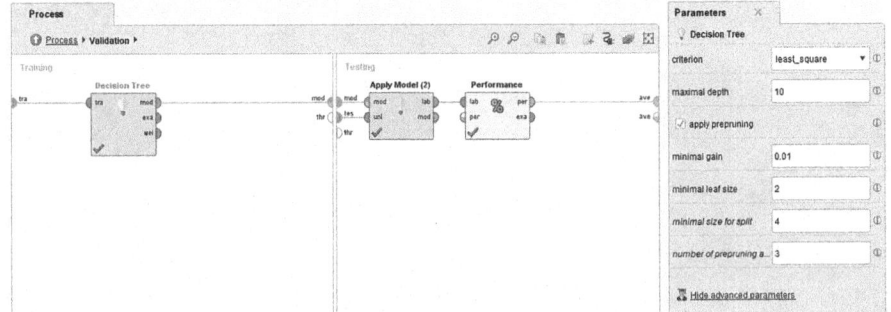

Figure 6-17b. *Regression tree inner process*

6. The predicted prices for the new data loaded are given in Figure 6-18.

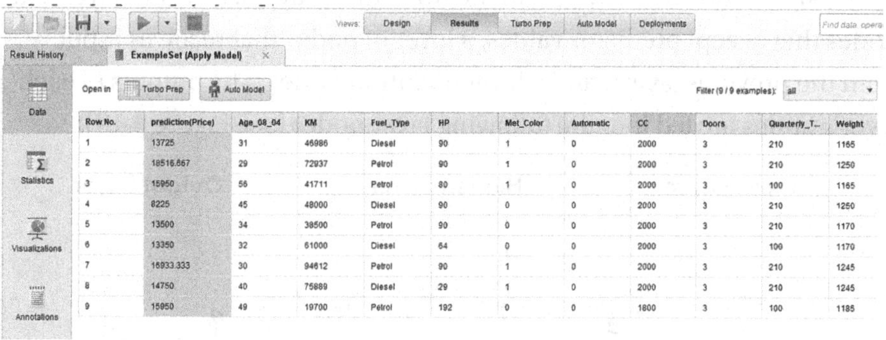

Row No.	prediction(Price)	Age_08_04	KM	Fuel_Type	HP	Met_Color	Automatic	CC	Doors	Quarterly_T...	Weight
1	13725	31	46986	Diesel	90	1	0	2000	3	210	1165
2	18516.667	29	72937	Petrol	90	1	0	2000	3	210	1250
3	15950	56	41711	Petrol	80	1	0	2000	3	100	1165
4	8225	45	48000	Diesel	90	0	0	2000	3	210	1250
5	13500	34	38500	Petrol	90	0	0	2000	3	210	1170
6	13350	32	61000	Diesel	64	0	0	2000	3	100	1170
7	16933.333	30	94612	Petrol	90	1	0	2000	3	210	1245
8	14750	40	75889	Diesel	29	1	0	2000	3	210	1245
9	15950	49	19700	Petrol	192	0	0	1800	3	100	1185

Figure 6-18. *Resulting datasets*

6.6 Neural Network (Prediction)

The neural network is a versatile data-driven method for classification and prediction. In terms of interpretability, it is referred to as a "black box." The principles of nodes and layers, as well as how they join to form the structure of a network, are used in neural networks. The predictive performance of a neural network is demonstrated by the fact that it captures the very complicated relationship between predictors and the target variable. It is quite tolerant of noisy data. Despite these benefits, neural networks lack a built-in variable selection mechanism, and their adaptability is strongly reliant on having enough data for training. When employing a neural network, it's important to keep an eye out for weights that lead to a local rather than a global optimum. Multilayer feedforward networks have been the most successful neural network applications in data mining.[4]

Multilayer feedforward networks

A feedforward network is a completely connected network with no cycles and a one-way flow. It is made up of three layers: an input layer with nodes that accept predictor values, a layer of nodes that receives input from the previous layers, and a layer of output nodes. The outputs of each layer's nodes are fed into the following layer nodes (see Figure 6-19).

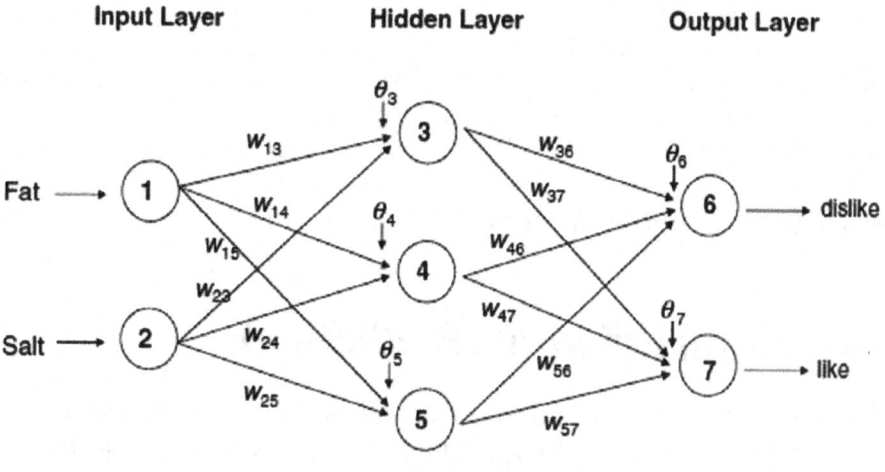

Figure 6-19. *Multilayer feedforward networks, Galit et al. (2018), p. 274*

Figure 6-19 shows an architectural design with two predictors, two hidden layers, and two nodes in the output layer that reflect the expected outcome value.

When we use a neural network for predicting numerical outcome, the predicted value needs to be converted to its original unit; this is because, to perform the prediction, both the predictors (numerical) and the outcome variables will have been normalized (rescaled to a [0,1] interval).

When working with neural networks, we must make the following decisions:

- *Network architecture*: This can be accomplished by trial-and-error runs on various architectures or through the use of algorithms.

- *Choice of predictor*: Depending on the software, you need to choose the learning rate and momentum. The learning rate has a range of 0 to 1 and is mostly used to avoid overfitting. The momentum is employed to keep the ball going while the weights are converged to the best possible position.

- *Number of hidden layers*: This is commonly set to one (though we can always experiment with others) because a single hidden layer is usually enough to capture even the most intricate interactions between predictors.

- *Size of hidden layer*: The level of intricacy of the relationship between the predictors that the network captures is determined by the number of nodes in the hidden layer. Starting with p (number of predictors) nodes and gradually decreasing or increasing while testing for overfitting is a good rule of thumb.

- *Number of output*: The number of nodes in a categorical outcome with m classes should be m or m-1. A single output node is usually used for a numerical result.

Moreover, to avoid overfitting in neural networks, the number of training iterations should be limited, and the data should not be overtrained. We look at the performance on the validation dataset or cross-validation dataset to see when it starts to deteriorate while the training set performance is still improving to detect overfitting.

Data preprocessing for neural networks: When the predictors and outcome variable are on a scale of [0,1], neural networks perform best. We normalize measurements for a numerical variable X that takes values in the range [a; b], by subtracting a from X and dividing by b - a. The normalized measurement is then

$$X_{norm} = \frac{X - a}{b - a}$$

(Eq. 6.9)

The minimal and maximal values of X in the data can be used to estimate a and b. Other than the creation of dummy variables, no adjustments are required for binary variables. If categorical variables with m categories are ordinal, choosing m fractions in the range [0,1] should represent their perceived ordering. If the categories are nominal, make m-1 dummies out of them. Note that it is also important to transform highly skewed predictors for better predictive performance. This situation is common with variables in the Business Analytics domain such as income. So it is advised to first transform this kind of variable by taking a log before scaling to [0,1].

Neural network – *evaluating performance (numerical outcome)*: The prediction performance of regression trees can be examined in the same manner that other predictive approaches (e.g., linear regression and regression trees) are evaluated, using summary measures like RMSE.

Sales prediction problem – neural network

Using the sales prediction problem introduced in Section 6.2, the following demonstrates how to use a neural network to solve the problem:

1. Create a new process, and drag the *ToyotaCorollaData2* data to the ***design view*** (remember that the ID column has been excluded in this version of the data). The overall process for running the neural network is given in Figure 6-20.

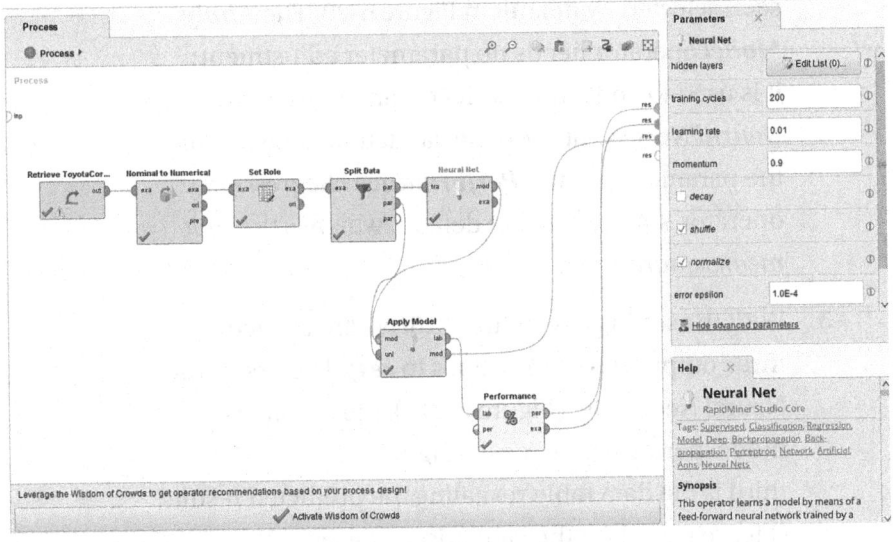

Figure 6-20. *Neural network process*

2. For the ***Nominal to Numerical*** operator, you
    will select the Fuel_Type attribute (converting
    Fuel_Type to dummies). Click the ***Set Role*** operator
    and set the ***attribute name*** to be *Price* and *target
    role* to be label. Click the ***Split Data*** operator, and
    for the ***partitions*** property of this data, add two
    entries, 0.6 and 0.4, respectively. For the ***Neural
    Net*** operator, we will leave the default setting as
    seen in Figure 6-20. This operator learns a model by
    means of a feedforward neural network trained by a
    backpropagation algorithm (multilayer perceptron).
    This operator cannot handle polynomial attributes
    (for more explanations on the parameters of the
    operator, check the help). Note that we did not use
    the normalize operator; this is taken care of by the
    normalize parameter of the ***Neural Net*** operator as

seen selected by default in Figure 6-20. The **Apply Model** operator needs no parameter adjustment; it is used to apply the model obtained from the **Neural Net** operator on the validation dataset. For the parameters of the **Performance (Regression)** operator, simply leave as default, which is the *root mean square error.*

3. With the settings in Figure 6-16, we are expecting three outputs (i.e., three links to **res**). The first is the Neural Net model (Figure 6-21a), the second is the model performance result (Figure 6-21b), and the third is the data table revealing the predicted results (Figure 6-21c) for the validation dataset.

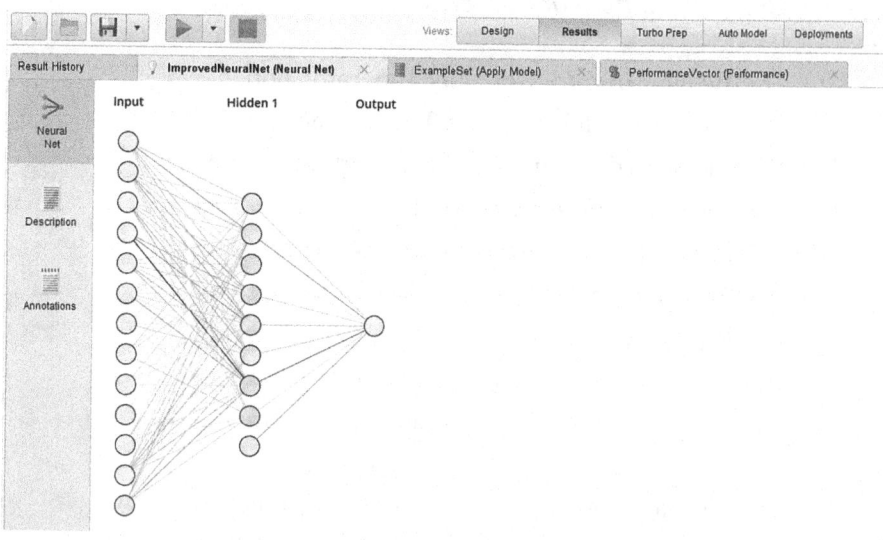

Figure 6-21a. *Neural network model*

The figure reveals the input, hidden, and output layers.

Figure 6-21b. *Evaluating the neural network*

From Figure 6-21b, we can see that the Neural Net
gives the best result out of the three algorithms
compared for this problem scenario.

Row No.	Price	prediction(Price)	Fuel_Type = ...	Fuel_Type = ...	Fuel_Type = ...	Age_08_04	KM	HP	Met_Color	Automatic	CC
1	13750	16328.446	1	0	0	23	72937	90	1	0	2000
2	13950	17063.663	1	0	0	24	41711	90	1	0	2000
3	18600	16169.130	1	0	0	30	75889	90	1	0	2000
4	21500	22254.378	0	1	0	27	19700	192	0	0	1800
5	12950	12800.799	1	0	0	23	71138	69	0	0	1900
6	20950	22359.171	0	1	0	25	31461	192	0	0	1800
7	19800	22341.172	0	1	0	25	32189	192	0	0	1800
8	21500	21850.915	0	1	0	31	23000	192	1	0	1800
9	22750	21787.076	0	1	0	30	34000	192	1	0	1800
10	17950	17203.617	0	1	0	24	21716	110	1	0	1600
11	16750	15496.592	0	1	0	24	25583	110	0	0	1600
12	16950	14967.670	0	1	0	30	64359	110	1	0	1600
13	16950	16824.541	0	1	0	29	43905	110	0	1	1600
14	16250	16219.704	0	1	0	29	25813	110	1	0	1600
15	12950	14629.263	0	1	0	29	9750	97	1	0	1400
16	15950	15322.558	0	1	0	27	29510	97	1	0	1400
17	14750	15110.612	0	1	0	27	27500	97	0	0	1400
18	13950	15563.728	0	1	0	22	49058	97	0	0	1400

ExampleSet (574 examples, 2 special attributes, 12 regular attributes)

Figure 6-21c. *Resulting dataset*

Figure 6-21c is the output of evaluating on the 40%
validation dataset and comparing the predicted
result and the original price in the validation
dataset.

4. For the Neural Net, we will use the settings in Figures 6-22a (main process) and 6-22b (inner process) to predict for new records.

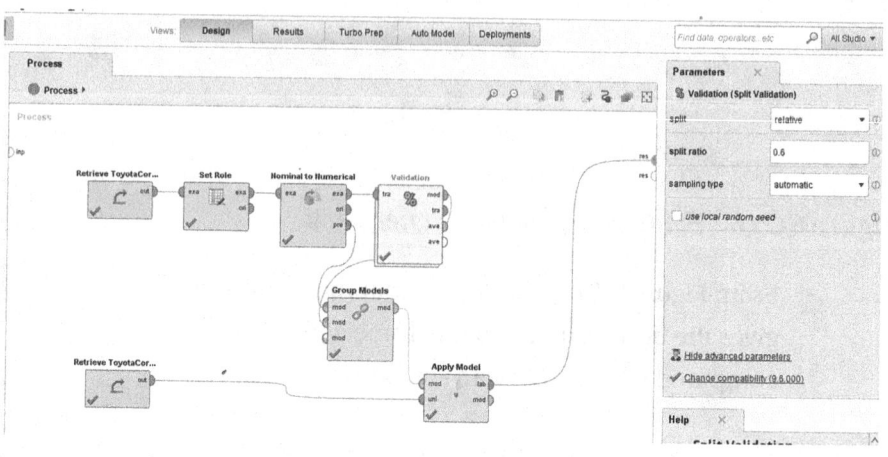

Figure 6-22a. *Predicting new data (Neural Net)*

5. Figure 6-22b is the inner process which you will create by double-clicking the ***Split Validation*** operator. The ***Neural Net, Apply Model,*** and ***Performance (Regression)*** operators are left as default settings just as earlier.

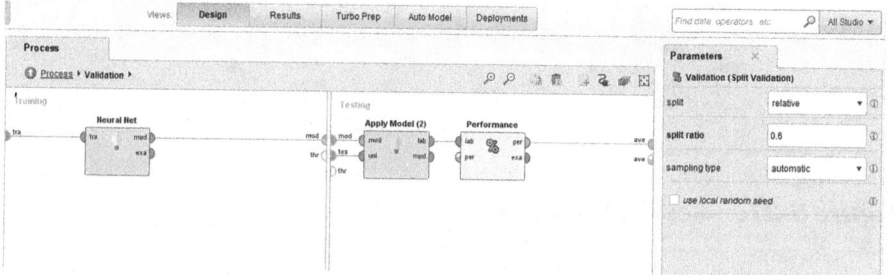

Figure 6-22b. *Inner process predicting new data*

6. The predicted prices for the new data loaded are given in Figure 6-23.

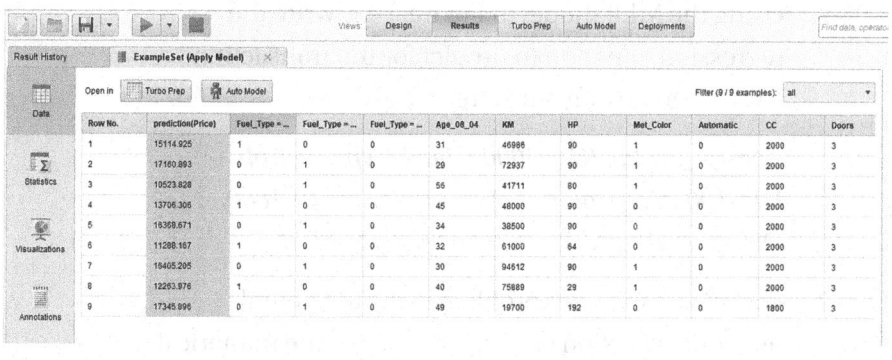

Figure 6-23. *Resulting dataset*

6.7 Conclusion on Sales Prediction

After trying several combinations of data attributes and algorithms, while checking for the best performance, there is a need to conclude which of the experiments you will use for your conclusion. In this case, we go with the algorithms that give the best prediction assuming we have been able to take care of overfitting and all. From Figure 6-24, we see that the algorithm giving the lowest error that is the lowest value of RMSE is the neural network.

MLR

root_mean_squared_error

root_mean_squared_error: 2949.808 +/- 0.000

RT

root_mean_squared_error

root_mean_squared_error: 1362.617 +/- 0.000

NN

root_mean_squared_error

root_mean_squared_error: 1307.861 +/- 0.000

Figure 6-24. *Comparing the results of the three algorithms*

In business applications, these models can be used in several ways, for example, it could be embedded in an application which gives the process of the Toyota Corolla in real time for making the decision concerning new customers and several others.

6.8 Problems

1. Using the MLR model in Figure 6-11, write the regression equation to predict for the transaction with the following attribute values:

    Age_08_04=29, KM=50000, Fuel_Type=Petrol, HP=100, Met_Color=0, Automatic=1, CC=1800, Doors=4, Quarterly_Tax=80, Weight=1170

2. For solving the sales problem using the neural network in Section 6.5, explore each of the numerical variables using the histogram, and for anyone that is skewed, first transform using the logarithm before normalization. Run the neural network model in Section 6.5 again and compare your results to the one in Section 6.5. Which one has a higher accuracy?

3. Given the data sample named *FoodAgricGroup.xlsx*, for each of the listed algorithm, create and evaluate models to predict Total Amount (target attribute). Which model gives the highest accuracy?

    a. Neural network

    b. Multiple linear regression

    c. Regression trees

Present the result obtained from each algorithm in detail for the following stages:

a. Data visualization

b. Feature selection

c. Model development

d. Model evaluation

6.9 References

1. Selva Prabhakaran (March 2017) Complete
   Introduction to Linear Regression in R, `www.`
   `machinelearningplus.com/machine-learning/`
   `complete-introduction-linear-regression-r/`

2. Jiawei Han, Micheline Kamber, and Jian Pei (2012)
   Data Mining Concepts and Techniques, Third
   Edition, published by Elsevier.

3. Galit Shmueli, Nitin R. Patel, & Peter C. Bruce,
   Data Mining for Business Analytics, Concepts,
   Techniques and Applications in R, published by
   John Wiley & Sons, Inc., Hoboken, New Jersey, 2018.

CHAPTER 7

Classification Techniques

In this chapter, even though there are several classification techniques, we will explore the popular ones used for classification in the business domain. In doing this, we will use the third business problem centered on customer loyalty using neural networks, classification trees, and random forest algorithms. In solving this problem, we are particular about how to get and retain more customers for our small business. We will also introduce some other classification-based techniques such as K-nearest neighbor and logistic regression. In using these techniques to solve the problem, we explain the fundamental concepts in the chosen algorithms and use them to demonstrate how these problem solving processes can be adopted in real business scenarios.

7.1 Classification Models and Evaluation

An important part of data analysis is the process of classifying the data that has been collected. Classification is a type of supervised machine learning approach that derives models used in this type of situations. This type of model is known as a classifier, and it is used to predict categorical (discrete, unordered) class labels. A typical example is to predict if a customer will leave a retail store. Classification models have been used

© Afolabi Ibukun Tolulope 2022
A. I. Tolulope, *Data Science and Analytics for SMEs*,
https://doi.org/10.1007/978-1-4842-8670-8_7

for several applications in business such as fraud detection, customer target marketing, staff performance prediction, and so on. Classification models work using a two-step process. The first is to construct the model from a labeled dataset (training data). The second is to use the model to classify future unknown objects (in the validation dataset, which is known but predicted using the newly developed model). In this step, we estimate the accuracy of the model by comparing the result with the original label (known) in the validation dataset. The validation dataset is independent of the training set; otherwise, overfitting will occur.

In classification, oversampling occurs when classes are found in very uneven ratios; random sampling may produce too few of the rare classes to yield relevant information about what differentiates them from the dominant class. To avoid these challenges, stratified sampling can be used. Stratified sampling is used to sample from data which can be partitioned into subportions.[2] The data is divided into homogeneous portions called strata (the plural of stratum).

We must judge classifiers because there are several types of classifiers and predicting approaches, and within each method, there are often multiple alternatives that can result in totally different outcomes. The probability of a *misclassification error* is used to evaluate the performance of classifiers. When a record is misclassified, it means it has been placed in a class by a model to which it does not belong.

The metrics for evaluating classifier performance are as follows:[1]

Accuracy and recognition rate

$$\frac{TN + TN}{P + N} \qquad \text{(Eq. 7.1)}$$

Error rate, misclassification rate

$$\frac{FN + FN}{P + N} \qquad \text{(Eq. 7.2)}$$

Sensitivity, true positive rate, recall

$$\frac{TP}{P}$$
(Eq. 7.3)

Specificity, true negative rate

$$\frac{TN}{N}$$
(Eq. 7.4)

Precision

$$\frac{TP}{TP + FP}$$
(Eq. 7.5)

F, F1, F-score, the harmonic mean of precision and recall

$$\frac{2 \ x \ precision \ x \ recall}{precision + recall}$$
(Eq. 7.6)

The terms true positive, true negative, false positive, positive, and negative samples are abbreviated as TP, TN, FP, P, and N, respectively.

- *True positives (TP)*: These are the positive tuples that the classifier properly labels.

- *True negatives (TN)*: These are the negative tuples that the classifier correctly categorized.

- *False positives (FP)*: These are negative tuples that have been mistakenly categorized as positive.

- *False negatives (FN)*: Positive tuples are classified as negative.

A confusion matrix is used to summarize the accurate and inaccurate classifications that a classifier generated for a given set of data, and this is done using the validation dataset for analyzing the performance of classifiers. In Table 7-1, the confusion matrix illustrates TP, TN, FP, and FN. Classifier performance can be assessed using a confusion matrix, which is a useful tool for testing how well a classifier detects tuples from different classes. A classifier's TP and TN show that it correctly classifies, whereas its FP and FN indicate that it incorrectly classifies, respectively. The Fbeta measure is a generalization of the F-measure that adds a configuration parameter called beta. Fbeta is used when both precision and recall are important but slightly more attention is needed on one or the other, such as when false negatives are more important than false positives, or the reverse.

Table 7-1. *Confusion Matrix*

Actual Class	Predicted Class			
		Yes	No	Total
	Yes	TP	FN	P
	No	FP	TN	N
	Total	P'	N'	P+N

Propensities and cut-off for classification

Some classification algorithms begin by calculating the probability that a record belongs to each of the classes. Another word for probability is "propensity." Propensities are frequently used as a first step in predicting class membership (classification) or in ranking records based on the probability of belonging to the desired class. To ensure that all classes are accurate, the record might be moved to a class that is most likely to get it (class with the greatest probability). Concentrate on a single class and compare the propensity (probability) of class membership to an analyst-determined cut-off value when there are many records and a single class is of interest. If the probability of inclusion in the desired class is greater than the cut-off, the record is moved to the desired class. There's a default cut-off value of 0.5 in two-class classifiers.[3]

7.2 Practical Business Problem III (Customer Loyalty)

Customer loyalty is described as the habit of repeatedly choosing a certain brand's or company's products or services notwithstanding the availability of a comparable product or service from another company, commonly referred to as a competitor. Loyal customers can never be swayed by the cost, pricing, or surplus of products from competitors. They would rather pay a higher price for the same high-quality service and product they are accustomed to. Starbucks credits its rewards program for the majority of the additional $2.65 billion in revenue. Customer loyalty benefits all types of organizations since it drives repeat business, boosts revenue, develops brand influencers and advocates, distinguishes a company from its competition, and provides honest and useful feedback.

In this third practical business problem, we will use and compare the results of the neural network, classification tree, and random forest using the same experimental conditions. The goal is to use a case study to demonstrate how we can use these algorithms to arrive at making inferences that can be used to get and retain customers for our businesses. In particular, we will predict the loyalty of bank customers and use this to know if a customer will be leaving the bank or not (i.e., how loyal is the customer). Based on the result of the prediction, an appropriate relationship marketing approach will then be used to encourage the customer to stay (if you predicted they will leave) or step up CRM for those that are loyal, for example, discounts for more sales. It's important to note that though we used a bank as a case study here, any other business can be used with appropriate data attributes.

The data for the business problem is named *Churn_Modelling.csv*, and it contains data on the customers of a bank. The data description is as follows:

• RowNumber	*Identification has to be removed*
• CustomerId	*Identification has to be removed*
• Surname	*Identification textual has to be removed*
• CreditScore	*Integer*
• Geography	*Categorical – indicating the country*
• Gender	*Categorical – indicating the sex*
• Age	*Integer*
• Tenure	*Integer*
• Balance	*Real*
• NumOfProducts	*Integer*
• HasCrCard	*Integer but has to be changed to categorical*
• IsActiveMember	*Integer but has to be changed to categorical*
• EstimatedSalary	*Real*
• Exited	*Outcome variable but has to be changed to categorical*

Regardless of the classification algorithm that will be used, the very first thing is to perform basic data preprocessing and explore the data using necessary visualization tools for the purpose of classification.

General data preprocessing for the classification techniques

1. Create a new *repository* and two *subfolders* (for data and for the process) in the repository. *Import*

the data named *Churn_Modelling.csv* into the data subfolder. When importing, remove RowNumber, CustomerID, and Surname columns and convert HasCrCard, IsActiveMember, and Exited to categorical attributes. Also, replace errors with missing values when importing the data. Figure 7-2 is the screenshot of importing the data.

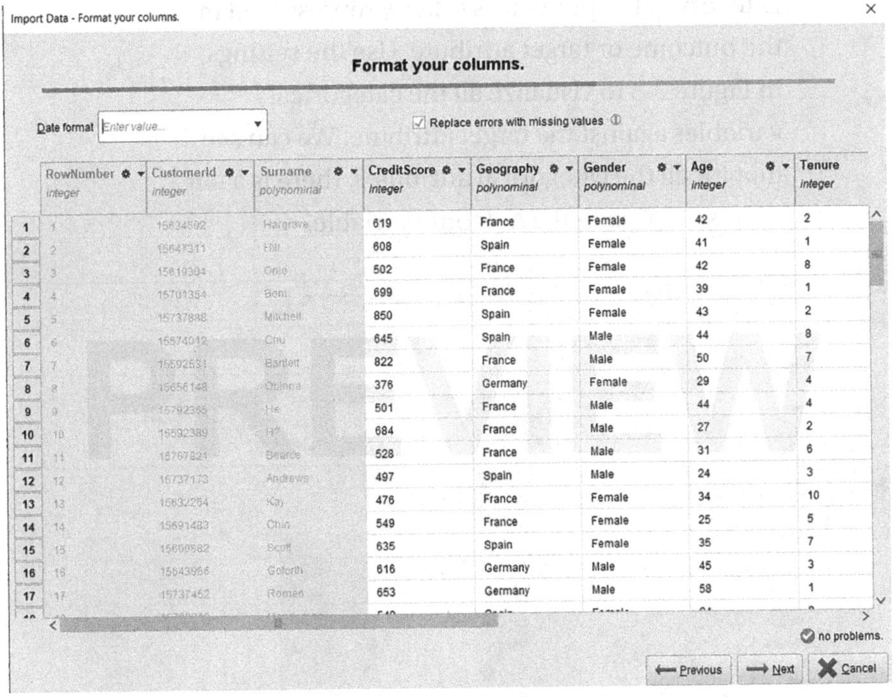

Figure 7-2.   *Importing the data*

2. After importing the data, drag it to the design view and connect it to **res**. Run the process and view the **statistics**. From the **statistics** view of the data, notice that the unnecessary attributes have been

removed, and the needed attribute change has occurred. Also, there are no missing values.

3. Using the guide for visualization, particularly with respect to classification (refer to Section 2.5 in Chapter 2), we will study the relation of outcome to categorical predictors using bar charts with the outcome on the y axis. The goal is to find out if the category of the predictors is fairly represented in the outcome or target attribute. Use the settings in Figure 7-3 to visualize all the categorical variables against the target attribute. We can see that for all the categorical attributes, there is a fair representation in the outcome variable.

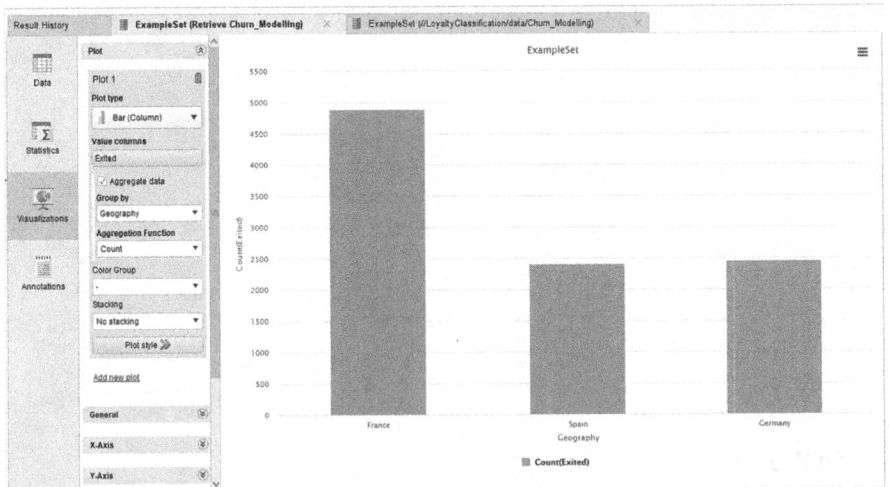

Figure 7-3. *Example of a bar chart of a categorical attribute and target variable*

4. Study the relation of outcome to pairs of numerical predictors via color-coded scatter plots (color denotes the outcome). The goal here is to see if there is a linear relationship between the numerical

predictors and to see the strength of the outcome
class in the numerical predictors. Use the settings
in Figure 7-4 to do this for different combinations of
the numerical attributes. Notice that there is no such
linear relationship. Incase there is such relationship,
it has to be handled because of multicollinearity.

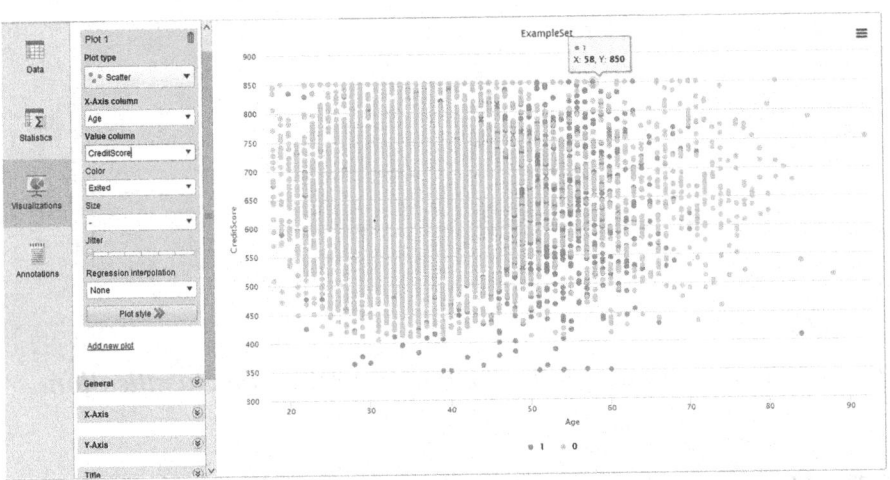

Figure 7-4. *An example of a color-coded scatter plot for classification*

5. Study the relation of outcome to numerical
predictors via side-by-side boxplots: plot boxplots
of a numerical variable by outcome. Create
similar displays for each numerical predictor.
The most separable boxes indicate potentially
useful predictors. Use the settings in Figure 7-5
to investigate this for all the numerical attributes.
Note that apart from NumOfProducts and
EstimatedSalary, all the other numerical attributes
are fairly separable as indicated by the red arrow
in Figure 7-5. (Their medians are fairly different).
In exercise 3 of this chapter, we will experiment the

classification process removing these attributes, but for now it will be included.

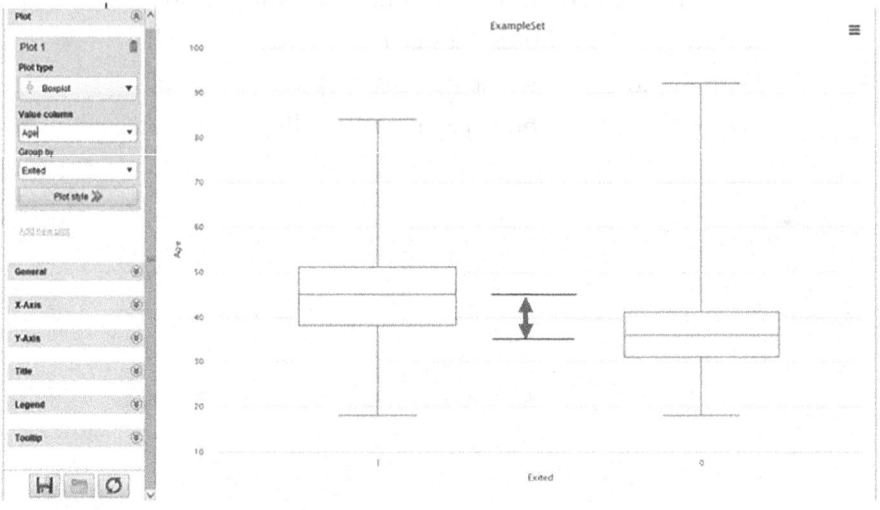

Figure 7-5. *An example of the boxplot visualized for classification*

7.3 Neural Network

Please refer to Section 6.5 for details on the neural network algorithm.

Customer loyalty with neural network

Using the customer loyalty problem introduced in Section 7.2, the following demonstrates how to use a neural network to solve the problem:

1. Create a new process, drag the already imported *Churn_Modelling* data to the design view. We will use the settings in Figures 7-6a (main process) and 7-6b (inner process) to develop and evaluate the model.

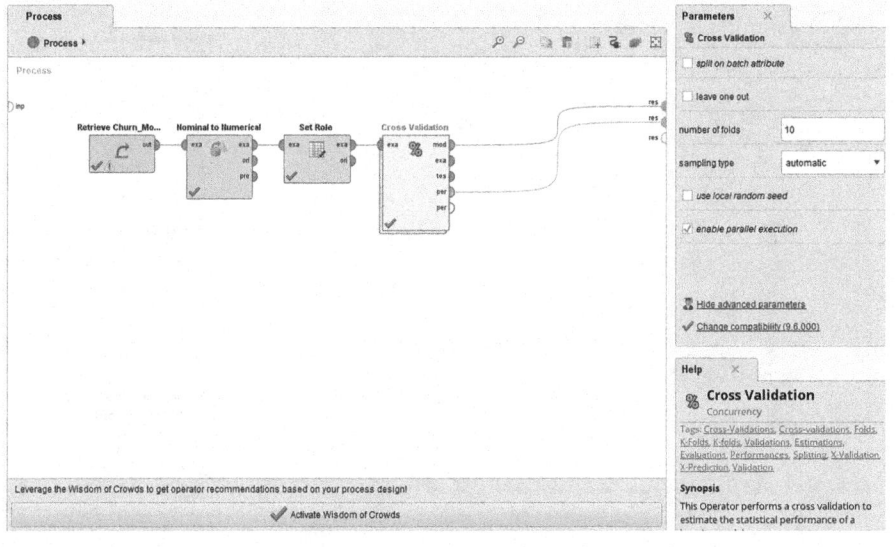

Figure 7-6a. *Main process of the neural network*

In Figure 7-6a, for the ***Nominal to Numerical***
operator you will select the attributes Gender,
Geography, HasCrCard, and IsActiveMember
(converting them to dummies). Click the ***Set Role***
operator and set the *attribute name* to be *Exited*
and the *target role* to be label. Leave the ***Cross
Validation*** operator as default settings, meaning
that we will be using a 10 folds validation.

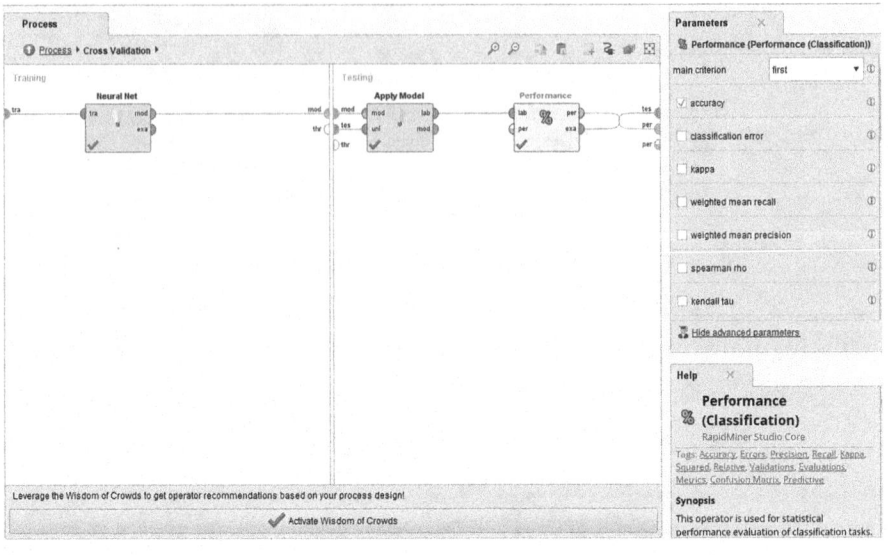

Figure 7-6b. *Inner processes of the neural network*

In Figure 7-6b for the **Neural Net** operator, we will click the hidden layer and give it any name (e.g., "ibk") and set it to −1. If the hidden layer size value is set to −1, the layer size would be calculated from the number of attributes of the input example set. More information in the RapidMiner documentation. All other settings remain as default. The **Apply Model** operator needs no parameter adjustment. For the parameters of the **Performance (Classification)** operator, simply leave as default, which is accuracy.

For the settings in Figures 7-6a and 7-6b, we are expecting two outputs. Figure 7-7a is the neural network model, and Figure 7-7b is the performance of the model.

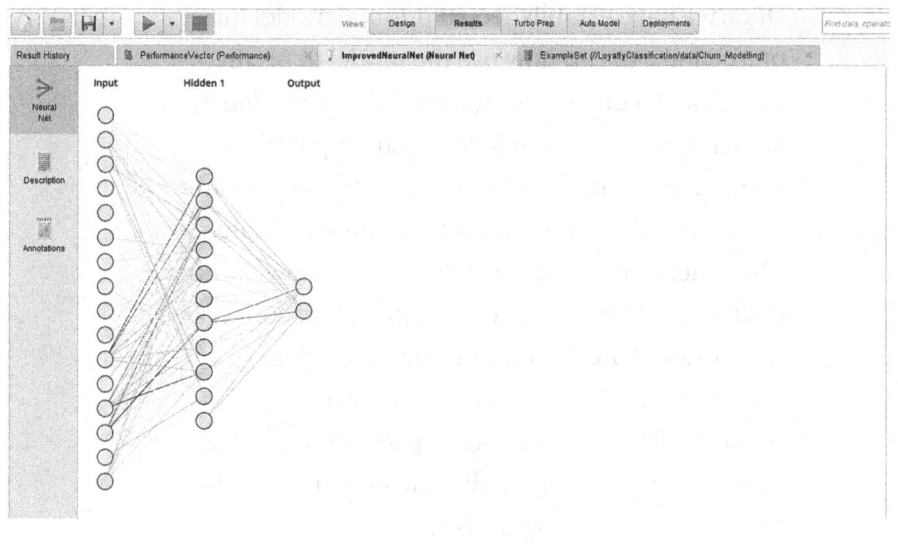

Figure 7-7a. *Neural network model*

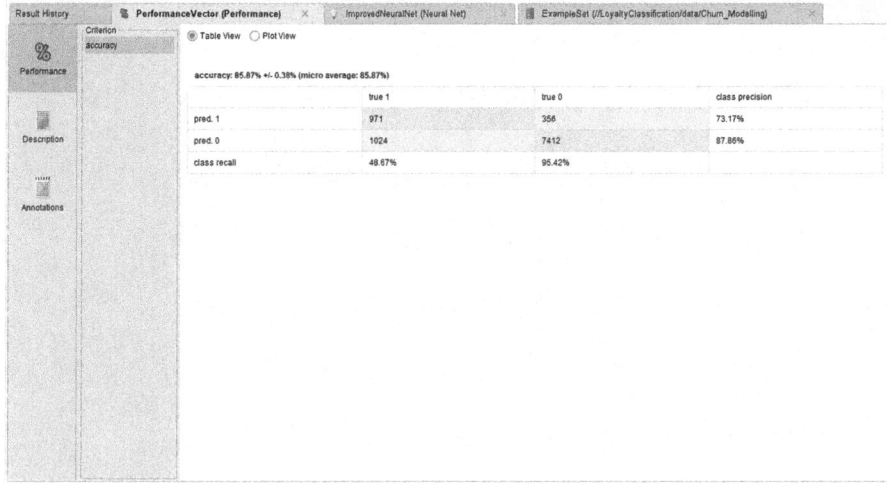

Figure 7-7b. *Performance result of the model*

2. Once we have established an optimal model for prediction, we can now use the model to predict new data. To do this, we will load the data *Churn_ModelingNew.xlsx* which has been prepared for predicting for new customers. The data is prepared such that it does not have the outcome variable. Also when importing the data, remove RowNumber, CustomerID, and Surname columns and convert HasCrCard, IsActiveMember, and Exited to categorical attributes. Also, replace errors with missing values when importing the data. To predict a new record, we will modify the main process in Figure 7-6a to give Figure 7-8.

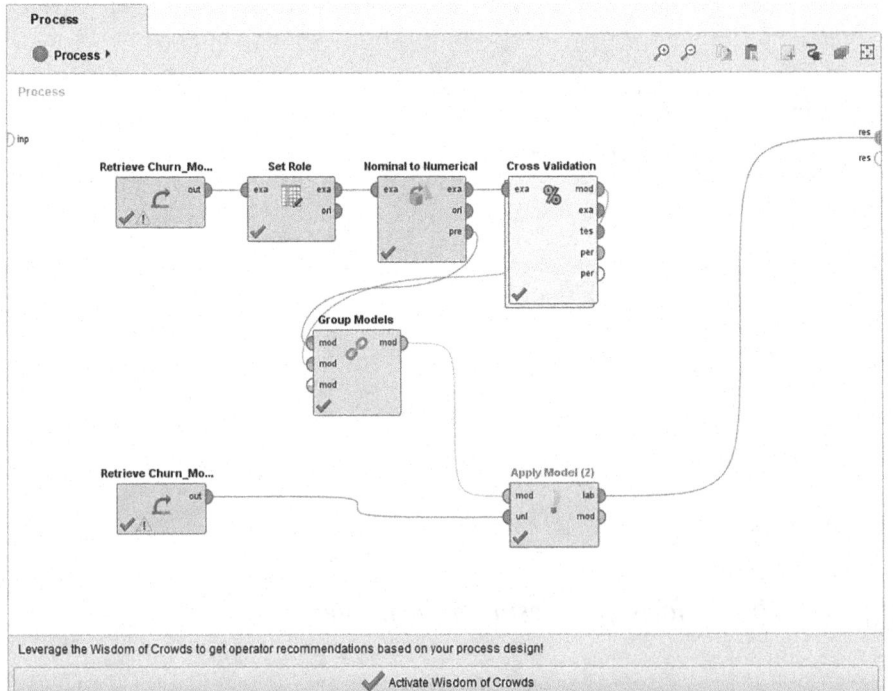

Figure 7-8. *Predicting for new records or transactions*

3. Figure 7-9 is the output from the modification in
    Figure 7-8. It shows the result for predicting the
    new transactions. It also reveals the confidence
    (propensity) of the predicted results. From these
    confidence results, we can see that the default
    threshold of 0.5 has been used to classify the
    customers into Exited (1) or not Exited (0).

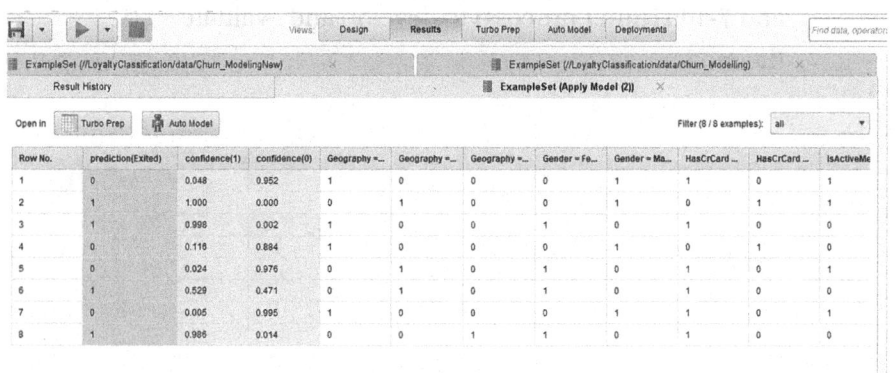

Figure 7-9. *Output of the new prediction*

7.4 Classification Tree

Please refer to Section 6.4 for details on decision tree algorithms.

Decision tree algorithms can be used to predict both numerical
and categorical outcomes. When the predicted outcome is categorical,
it is referred to as classification trees. Classification trees use predictor
information to "drop" a record down the tree until it reaches a node at the
bottom (terminal node). We can then classify it by gathering a "vote" of all
the training data that belongs to the terminal node during the formation of
the tree. The class with the most votes gets the new record.

Customer loyalty with classification tree

Using the customer loyalty problem introduced in Section 7.2, the following demonstrates how to use the classification tree to solve the problem:

1. Create a new process, and drag the already imported *Churn_Modelling* data to the design view. We will use the settings in Figures 7-10a (main process) and 7-10b (inner process) to develop and evaluate the model.

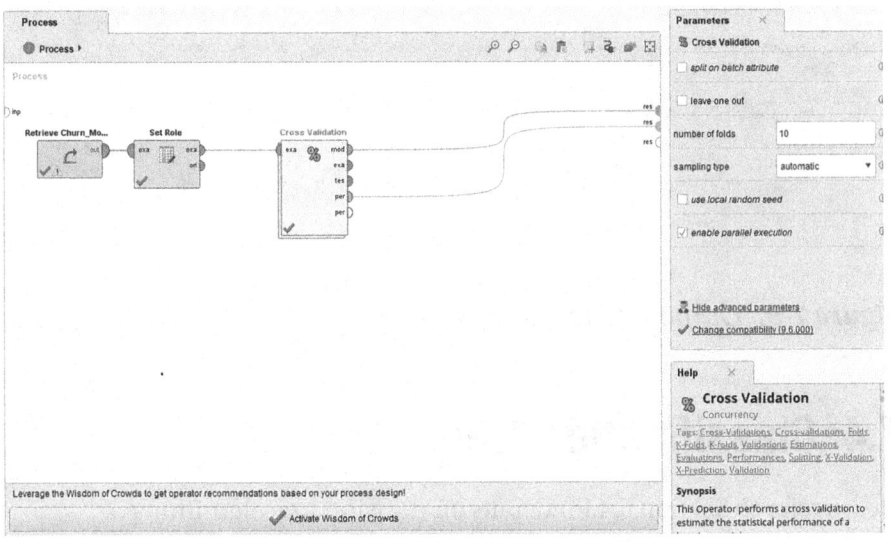

Figure 7-10a. *Decision tree main process*

The ***Decision Tree*** operator generates a decision tree model, which can be used for classification and regression. This operator can process ExampleSets containing both nominal and numerical attributes. The label Attribute must be nominal for classification and numerical for regression.

In Figure 7-10a, click the **Set Role** operator and set
the *attribute name* to be *Exited* and *target role* to
be label. Leave the **Cross Validation** operator as
default settings, meaning that we will be using a
tenfold validation.

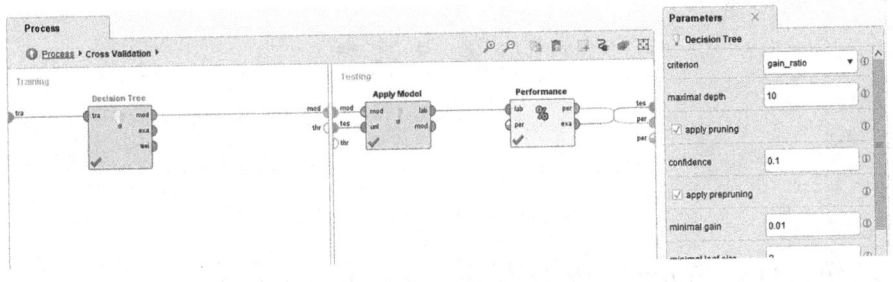

Figure 7-10b. *Inner process of the decision tree*

In Figure 7-10b for the **Decision Tree** operator,
we set the criterion to gain_ratio. This signifies
that we are predicting categorical outcome.
The **Apply Model** operator needs no parameter
adjustment. For the parameters of the **Performance
(Classification)** operator, simply leave as default,
which is accuracy.

For the settings in Figures 7-10a and 7-10b, we
are expecting two outputs. Figure 7-11a is the
neural network model, and Figure 7-11b is the
performance of the model.

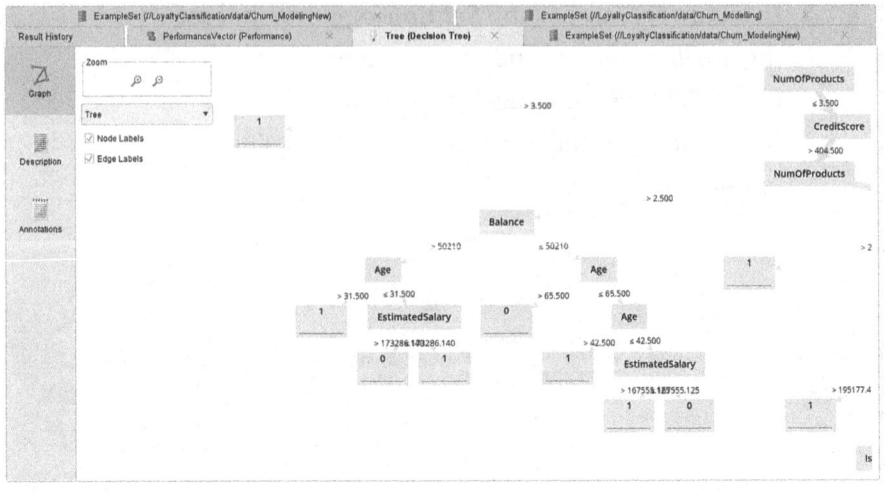

Figure 7-11a. *Some sections of the decision tree model*

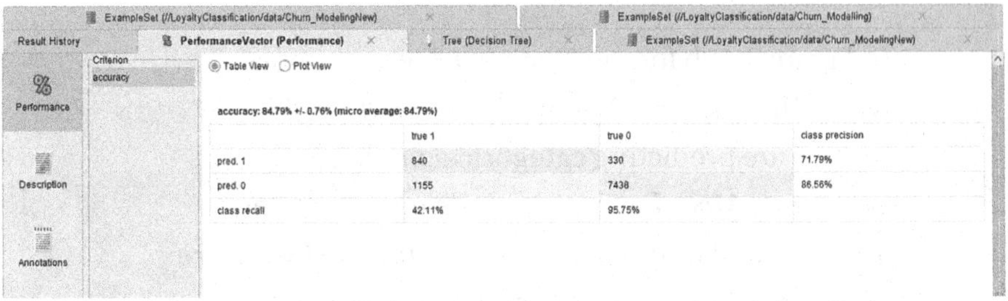

Figure 7-11b. *Performance of the decision tree model*

2. Once we have established an optimal model for prediction, we can now use the model to predict for new data. To do this, we will load the data *Churn_ModelingNew.xlsx* which has been prepared for predicting new customers. To predict a new record, we will modify the main process in Figure 7-10a to give Figure 7-12.

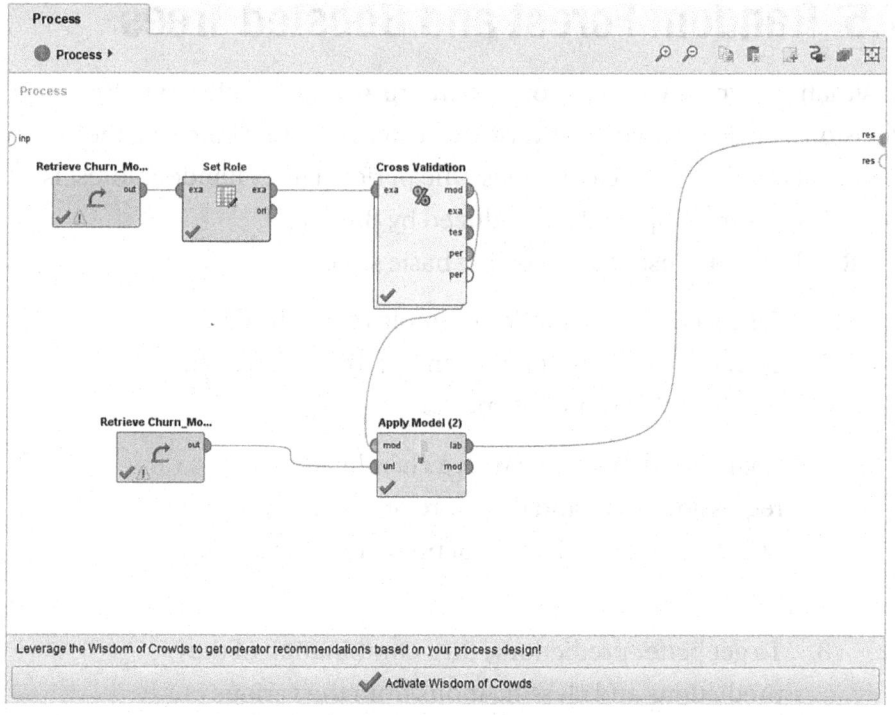

Figure 7-12. *Predicting for new records (decision trees)*

3. Figure 7-13 is the output from the modification in
    Figure 7-12. It shows the result for predicting the
    new transactions.

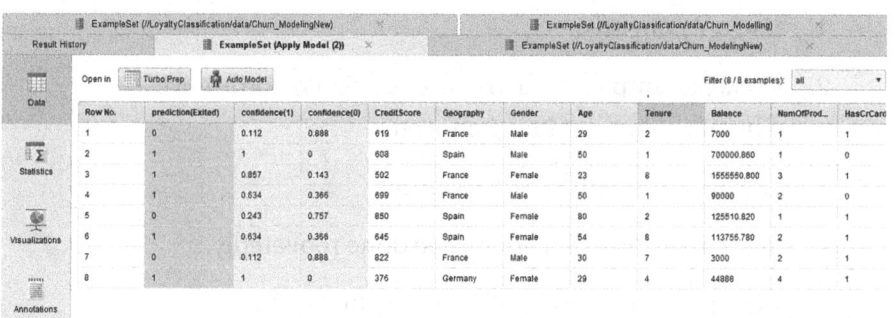

Figure 7-13. *Predicted records (decision trees)*

173

7.5 Random Forest and Boosted Trees

In situations where visualization and interpretation of rules from the decision tree do not matter, several tree extensions that combine the results of multiple trees can increase the performance. Random forest is a popular multitree approach, introduced by Breiman.[5]

Random forest uses the following basic steps:

1. Take a number of random samples from the data, each with a different replacement. Bootstrap is the name for this sampling method.

2. Adapt/fit each sample with a classification (or regression) tree (and therefore produce a "forest") using a random selection of predicting variables at every stage.

3. To get better predictions, do a combination of the predictions and classifications from the various trees. Voting is used for classification, while averaging is used for prediction.

Boosted trees are another type of multitree improvement. A sequence of trees is fitted, with each tree focusing on records from the preceding tree that were incorrectly classified. It has the following steps:

1. Use only one tree.

2. Create a sample in which misclassified records have a higher chance of being chosen.

3. Create a tree with the new sample.

4. Steps 1 and 2 earlier should be done repeatedly.

5. Classify records using weighted voting, with subsequent trees receiving more weight.

If we run a boosted tree on the same data, it slightly outperforms the single pruned tree. In terms of overall accuracy, the boosted tree outperforms the validation data, particularly when it comes to correctly classifying 1s (rare classes of special interest).

Customer loyalty with random forest

1. Create a new process, and drag the already imported *Churn_Modelling.xlsx* data to the design view. We will use the settings in Figure 7-14a (main process) and 7-14b (inner process) to develop and evaluate the model.

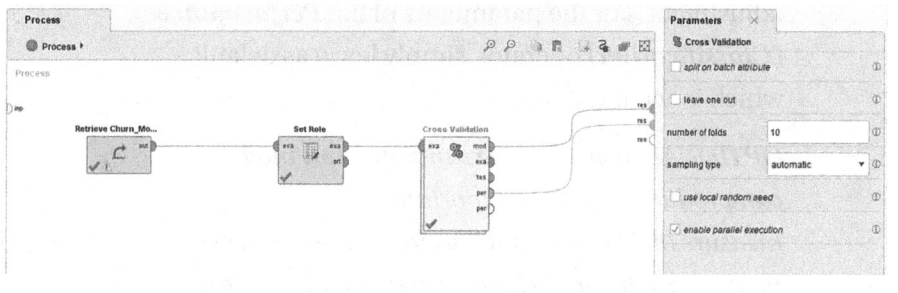

Figure 7-14a. *Main process of random forest*

The **Random Forest** operator generates a model which is a combination of several trees. We will leave the default which is 100 trees as seen in its *Number of Trees* parameter. In Figure 7-14a, click the *Set Role* operator and set the *attribute name* to be *Exited* and the *target role* to be *label*. Leave the **Cross Validation** operator as default settings, meaning that we will be using a tenfold validation.

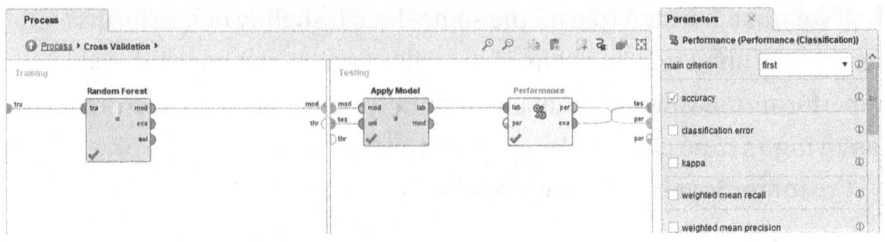

Figure 7-14b. *Inner process for the random forest*

In Figure 7-14b for the ***Random Forest*** operator, we leave the number of trees as default. The ***Apply Model*** operator needs no parameter adjustment. For the parameters of the ***Performance (Classification)*** operator, simply leave as default, which is accuracy.

*OPTIONAL: Note that the random forest model can produce "variable importance" scores, which measure the relative contribution of the different predictors. The importance score for a particular predictor is computed by summing up the decrease in the Gini index for that predictor over all the trees in the forest. (To get this result, alter the **Random Forest** operator in the inner process (i.e., Figure 7-14b) by right-clicking it and putting the breakpoint after, then connect the **wei** to **thr**. If you run the process, you will see the strength of each attribute.)*

2. For the settings in Figures 7-14a and 7-14b, we are expecting two outputs. Figure 7-15a is the random forest model, and Figure 7-15b is the performance of the model.

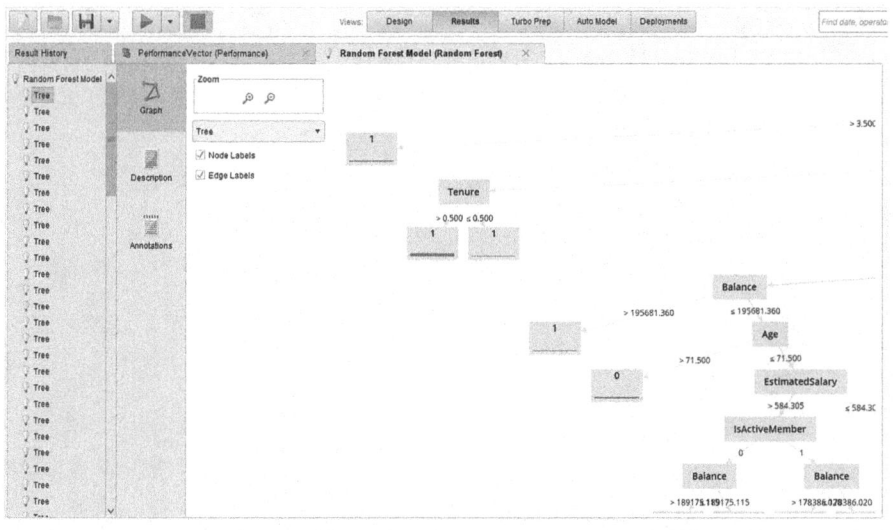

Figure 7-15a. *Some sections of the random forest model*

The list of the trees is displayed in Figure 7-15a. If you click each tree, you will see the model of the tree as shown in Figure 7-15a. There are a total of 100 different trees.

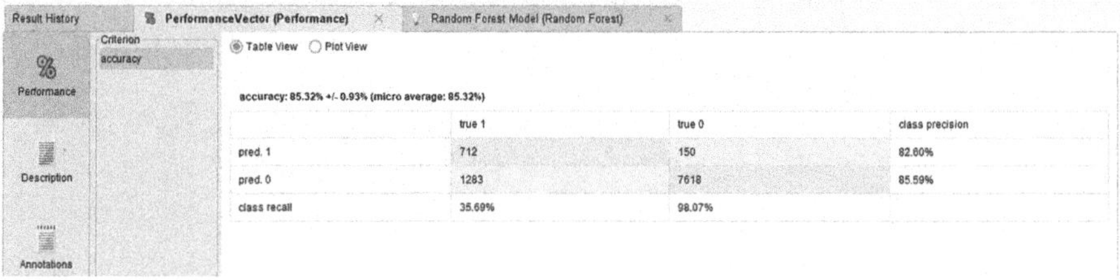

Figure 7-15b. *Performance of the random forest model*

For customer loyalty with boosted trees

To run the boosted tree algorithm, use the same settings for the random forest in Figures 7-14a and 7-14b except that in Figure 7-14b the ***Random Forest*** operator will be replaced with the ***Gradient Boosted Trees*** operator. The resulting model and performance are displayed in Figures 7-16a and 7-16b, respectively.

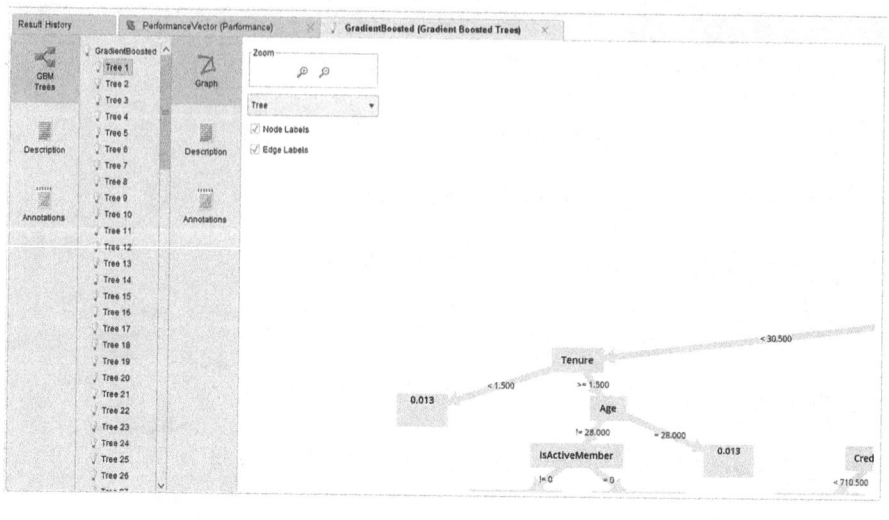

Figure 7-16a. *Some sections of the boosted trees*

The list of the trees is displayed in Figure 7-16a. If you click each tree, you will see the model of the tree as shown in Figure 7-16a. There are a total of 100 different trees.

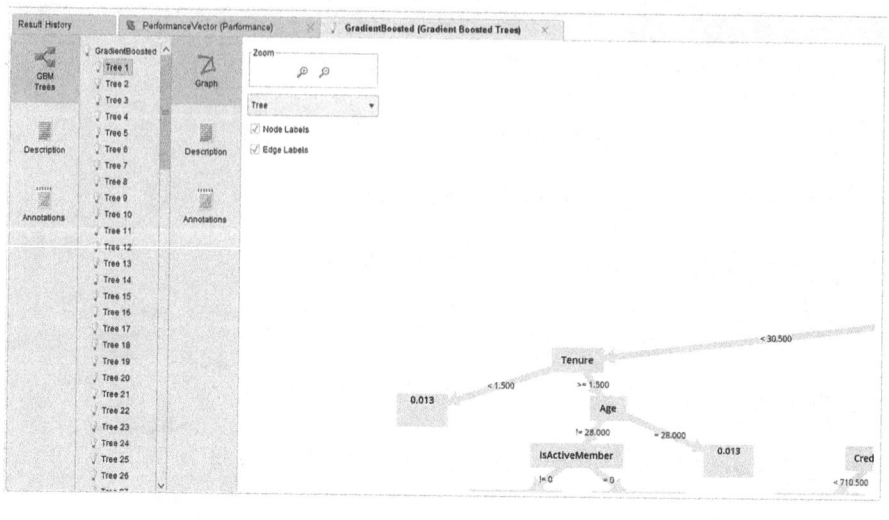

Figure 7-16b. *Performance of the boosted tree model*

Having used the different algorithms to solve this problem, we can now compare their accuracy and select the best for this particular data. Just as for the prediction, for the classification too, there might be a need to use different data and algorithm setting to experiment for better accuracy.

Comparing Figures 7-7b, 7-11b, 7-15b, and 7-16b, we can see that the best of the algorithms is the neural network (Figure 7-7b), which gave an accuracy of 85.79%, just slightly higher than random forest.

For this problem, the objective is to predict which customers will be loyal to prospective customers. With the result of the prediction for the new customers, we can identify this, particularly if we select those that have a high propensity. Also for the existing customers, we can predict the loyal ones and the ones about to churn and take necessary actions. If, for example, the goal is to target prospective customers, we can rank the prospective customer records from highest to lowest propensity (in the newly predicted dataset) in order to act on those with the highest propensity for target marketing.

7.6 K-Nearest Neighbor

In this section, we'll look at the K-nearest neighbor (KNN) algorithm and how it can help us obtain more clients using a practical example. K-nearest neighbor can be used to predict both numerical and categorical output. In this example, it will be used to predict the categorical output. This is particularly because the task of getting more customers is a classification task.

KNN searches the training dataset for records that are "similar" for the purpose of classifying or predicting a new record. By using averaging for prediction as well as voting for classification, these "neighbors" are used to produce a classification or prediction for the new record. The method is to search the training data for records that are comparable to (or have values similar to) the record we want to classify. We classify the one we wish to classify based on the classes that these related records fall under.

How do we determine these similar records (the neighbors)?

The Euclidean distance is a distance measurement that is extensively used. Between two records $(x_1, x_2........x_p)$ and $(u_1, u_2....... u_p)$, the Euclidean distance is

$$\sqrt{\left(x_1 - u_1\right)^2 + \left(x_2 - u_2\right)^2 + ... + \left(x_p - u_p\right)^2} \qquad \text{(Eq. 7.7)}$$

- Prior to computing a Euclidean distance, in most circumstances, the predictors should be normalized so that the scales of each individual predictor are equal.

- It's also worth noting that new records are standardized using the mean and standard deviations from the training data, and the new record is not included in calculating them.

- Validation data, like new data, is excluded from this calculation.

To classify, KNN uses the following rules:

- If $k = 1$ (the simplest case)

- We find the closest record (the nearest neighbor) and classify the new record as being in the same class as its closest neighbor.

- If $k > 1$

- Locate the k neighbors that are closest to the one being classified.

- Classify the record using a majority decision rule, with the record being classified as a member of the k neighbors' majority class.

In business, KNN can be used in most classification scenarios and works well when the available data is not much.

It is crucial to reduce the amount of time to calculate distances while preprocessing data for the KNN algorithm by working with fewer attributes. This is because the anticipated distance to the nearest neighbor increases drastically with the number of predictors. To reduce the attributes, we can apply different dimension reduction techniques. Also, the number of records necessary in the training set to qualify for KNN increases exponentially with the number of predictors. As a result, methods such as selecting subsets of predictors for our model or merging them using methods like principal component analysis, singular value decomposition, and factor analysis are needed to reduce the number of predictors.[3]

Choosing K

To avoid overfitting owing to noise in the training data, the k value chosen should be bigger than 1. Because it can suit the noise in the data, K should not be too low. K should not be set too high, as this will prevent the algorithm from capturing the data's local structure. The optimal option of k is determined by the data's nature. The range of k values should be between 1 and 20. To avoid ties, we'll use odd numbers. The best categorization performance determines the value of K. We classify the records in the validation data using the training data, then compute error rates for different k values. In RapidMiner, select k to be a small (between 1 and 14), positive, and odd integer. Select several k and run the following process iteratively for those k values till you get the one with the best accuracy.

Also, categorical variables must be transformed to binary dummies before KNN can be employed, and unlike statistical models such as regression, all binaries must be created and used with KNN. It's also important to normalize the features (due to the use of Euclidean distance, for example).

For KNN, the data partitioning is advised as follows:

- *Training*: Develop the models for different values of k, and then select the best value of k.

- *Validation*: To be used to evaluate the accuracy for each value of k using the training data and then select the best value of k.

- *Testing*: When the best value of k is selected, then this is used to check the final accuracy.

The evaluation of the KNN model can be done based on the methods for evaluating prediction or classification as the case demand since it can be used for both.

Classification example using KNN

A picture frame designer would like to find a way of classifying their customers in Lagos into those likely to purchase a family size frame and those not likely to buy one. We will use a random sample of 12 families who have purchased the frames and 12 who have not, living in Ibadan (a similar city).

1. The data used for this example is called *FramesData.xlsx*. Create a new **repository** and two **subfolders** (data and process). Create a new process. **Import** the data and check the **Statistics**. From the statistics, there are no missing values.

2. Visualization: Study the relation of outcome to numerical predictors via side-by-side boxplots. The most separable boxes indicate potentially useful predictors. Figures 7-17a and 7-17b show a fair result in this regard.

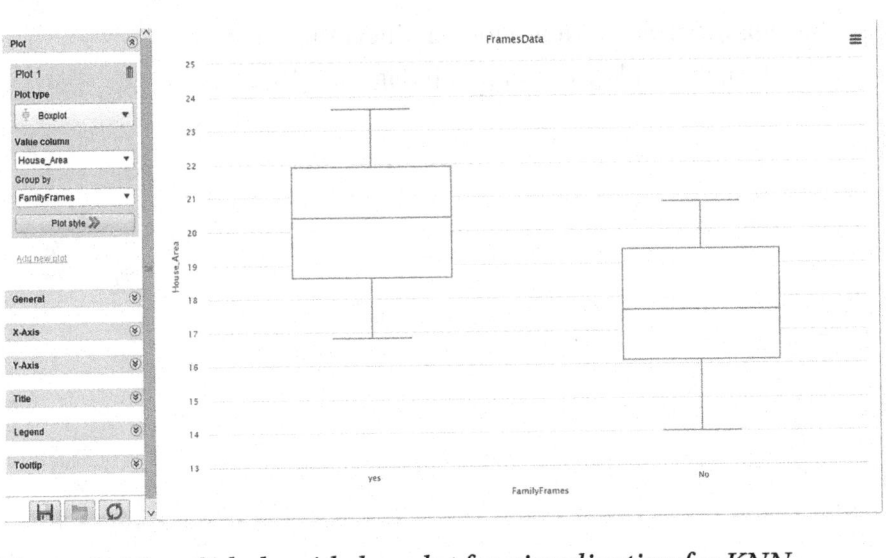

Figure 7-17a. *Side-by-side boxplot for visualization for KNN (House Area)*

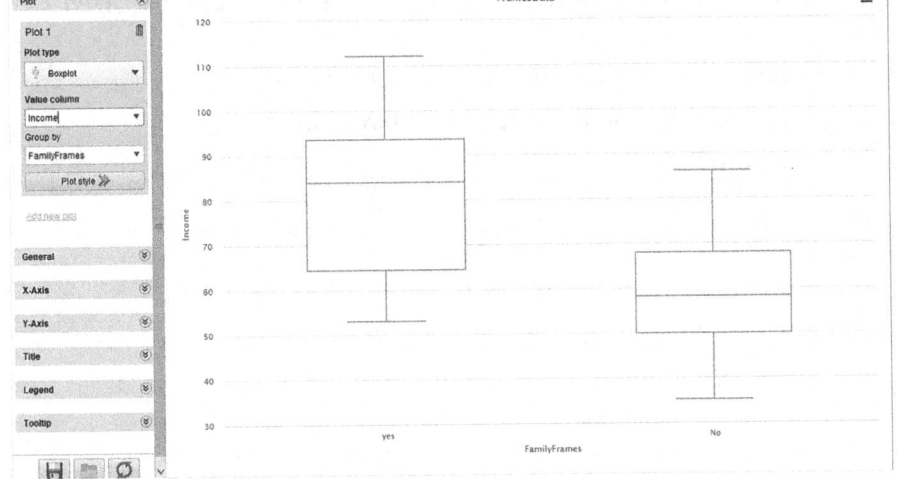

Figure 7-17b. *Side-by-side boxplot for visualization for KNN (Income)*

3. For more visualization for classification, we will use
    the color-coded scatter plot in Figure 7-18.

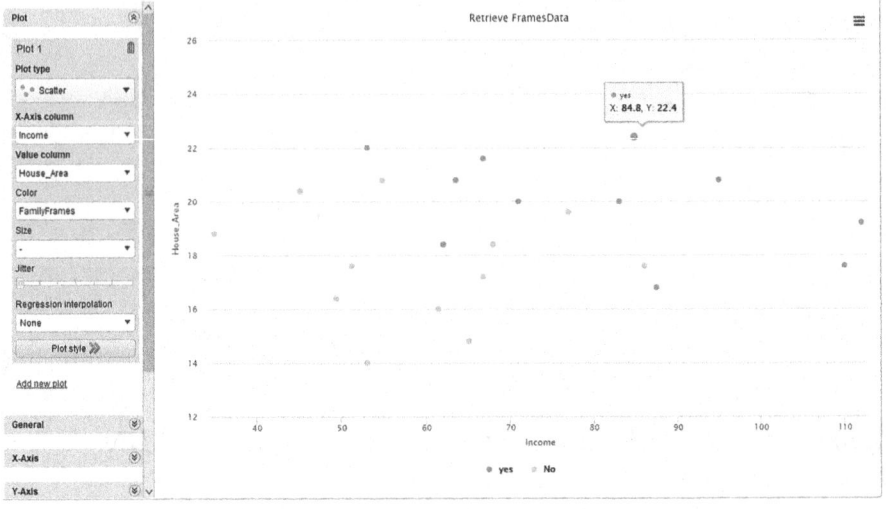

Figure 7-18. *Color-coded scatter plot for Income and House Area*

Use the settings in Figure 7-19 to check the
correlation, and the result is displayed in
Figure 7-20.

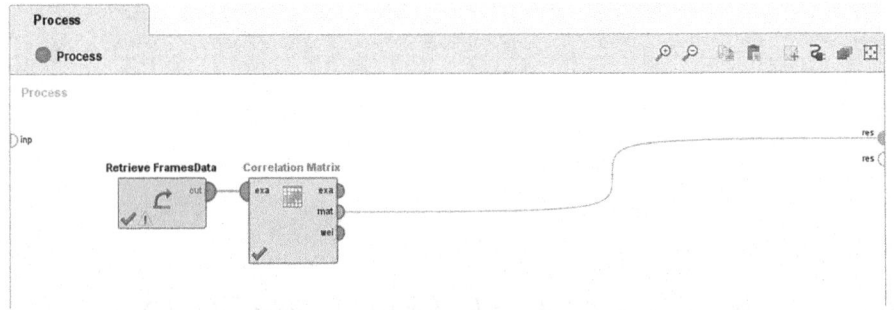

Figure 7-19. *Correlation matrix process for FramesData.xlsx*

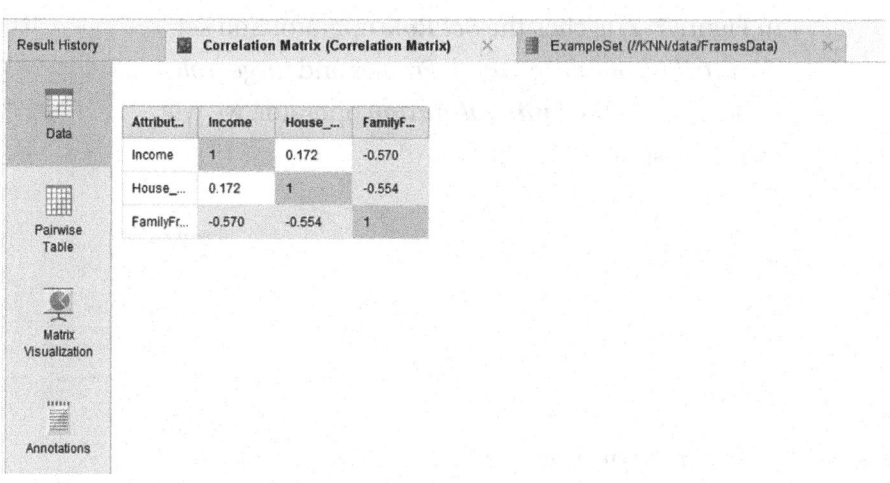

Figure 7-20.   *Correlation matrix output for FramesData.xlsx*

4.   We will use the settings in Figures 7-21a (main
     process) and 7-21b (inner process) to develop and
     evaluate the model. For this simple demonstration,
     because the data is not much, we will divide into
     only two partitions and use the validation data
     to determine the k and also to test the model
     obtained from the training data for the purpose of
     demonstration. In real business application, the
     data used to test the model after k has been selected
     and should be the third partition which has not
     been used at all.

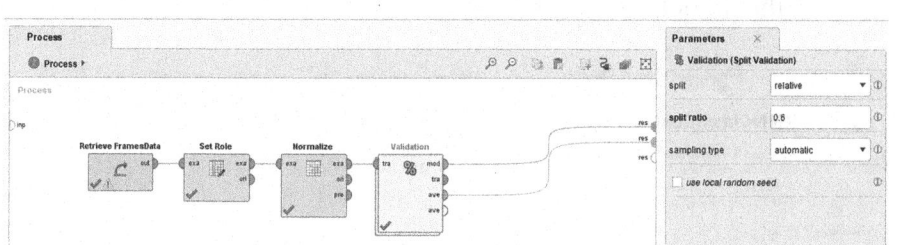

Figure 7-21a.   *Main process of KNN*

In Figure 7-21a, click the **Set Role** operator and set the *attribute name* to *FamilyFrames* and *target role* to be label. In the **Split Validation** operator, we will make the split ratio to be 0.6.

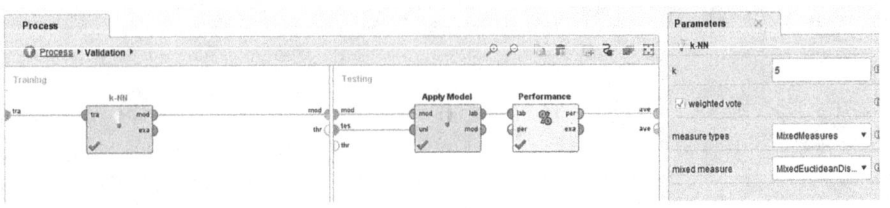

Figure 7-21b. *Inner process of KNN*

In Figure 7-21b for the **KNN** operator, we set the number of k to 5 (default). (It is advised to run KNN and evaluate the KNN model on values of K = 3,7,9,11,13, and record the performance, i.e., accuracy, to determine the best k.) All other settings remain as default. The **Apply Model** operator needs no parameter adjustment. For the parameters of the **Performance (Classification)** operator, simply leave as default, which is accuracy.

For the settings in Figures 7-21a and 7-21b, we are expecting two outputs. Figure 7-22a is the KNN model, and Figure 7-22b is the performance of the model.

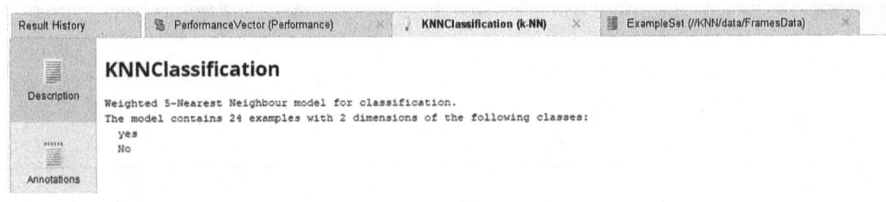

Figure 7-22a. *KNN model*

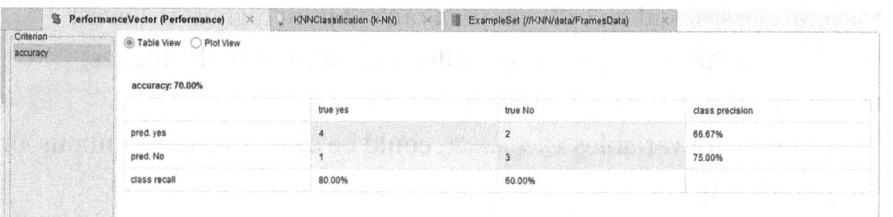

	true yes	true No	class precision
pred. yes	4	2	66.67%
pred. No	1	3	75.00%
class recall	80.00%	60.00%	

accuracy: 70.00%

Figure 7-22b. *Performance of the model*

5. Once we have established an optimal model for prediction and the best k that gave this model, we can now use the model to predict for new data. The procedure is similar to the already explained classification algorithms.

7.7 Logistic Regression

Logistic regression is based on a model that links the predictors to the result. A small set of data can be used to train logistic regression classifiers, which can then be applied to large groups of new records. When the result variable, Y, is categorical, logistic regression extends the concepts of linear regression.

Based on the values of its predictor variables, logistic regression may be used to classify a new record, whose class is not known, into one of the classes (called classification). It can also be used to discover features distinguishing between records in various classes in terms of their predictor profile or predicting variables in data where the class is known (called profiling).

When we need to classify customers as returning or nonreturning (classification) in business, find features that discriminate between customers (profiling), and so on, we use logistic regression. We can use logistic regression to calculate a new record's propensity to fall into one of several classes, and records are then ranked from highest to lowest propensity so that we can focus our efforts on those with the highest propensity.

The predictor variables x_1, x_2,......x_p could be categorical, continuous, or a combination of both.

Steps of logistic regression

1. Calculate the propensities of belonging to each class.

2. Use a cut-off value on the propensities to classify each record into one of the classes. The record is assigned to the class with the highest probability.

Just like in KNN, the default cut-off value in two-class classifiers is 0.5.

The logistic regression model

From the linear regression equation in Equation 6.7, we can express p as a linear function *predictor* $(x_1, x_2......x_p)$:

$$Y = \beta_0 + \beta_1 X_1 + \beta_2 X_2 + \beta_3 X_3 + \ldots\ldots\ldots + \beta_p X_p \qquad \text{(Eq. 7.8)}$$

Instead, the logistic regression uses a nonlinear function of the predictors in the form

$$p = \frac{1}{1 + e^{-(\beta_0 + \beta_1 x_1 + \beta_2 \beta x_2 + \ldots + \beta_q \beta x_q)}} \qquad \text{(Eq. 7.9)}$$

This is what is referred to as the logistic response function. For any value $x_1...x_q$, the right-hand side will always result in values between [0; 1].

Odds are a measure of belonging to a specific class. The ratio of the probability of belonging to class 1 to the probability of belonging to class 0 is defined as the odds of belonging to class 1. We may also do a computation in the opposite direction. We can compute the probability of an event using Equation 7.9 if we know the odds. If we compute the probability in the opposite direction, we get the following result:

$$Odds(Y = 1) = \frac{p}{1 - p} \qquad \text{(Eq. 7.10)}$$

If the chance of winning is 0.5, for example, the odds of winning are 0:5/0:5 = 1, that is, the probability of being in class 1 divided by the probability of being in class 0. We can perform the reverse calculation; we can compute the probability of an event if we know the odds by modifying Equation 7.10. This gives us Equation 7.11.

$$p = \frac{odds}{1 + odds} \qquad \text{(Eq. 7.11)}$$

We may describe the relationship between the odds and the predictors by substituting, for example, Equation 7.9 into Equation 7.11 as

$$Odds(Y = 1) = e^{\beta_0 + \beta_1 x_1 + \beta_2 x_2 + \ldots + \beta_q x_q} \qquad \text{(Eq. 7.12)}$$

Now, we can derive the standard formulation of a logistic model by using a natural logarithm[2] on both sides:

$$\log(odds) = \beta_0 + \beta_1 X_1 + \beta_2 X_2 + \beta_3 X_3 + \ldots\ldots + \beta_p X_p \qquad \text{(Eq. 7.13)}$$

For more information, visit Galit et al. (2018), Chapter 10.

Classification example using logistic regression

The dataset used for this example is named UniversalBank.xlsx. It consists of data on 5000 customers. The data include the customer's response to the last personal loan campaign (personal loan), as well as customer demographic information (age, income, etc.) and the customer's relationship with the bank (mortgage, securities account, etc.). The details are as follows:

• Age	*Customer's age in completed years*
• Experience	*Number of years of professional experience*
• Income	*Annual income of the customer ($000s)*
• Family	*Family size of the customer*
• CCAvg	*Average spending on credit cards per month ($000s)*
• Education	*Education level. 1: Undergrad; 2: Graduate; 3: Advanced/ Professional*
• Mortgage	*Value of house mortgage if any ($000s)*
• Securities Account	*Coded as 1 if customer has securities account with bank*
• CD Account	*Coded as 1 if customer has certificate of deposit (CD) account with bank*
• Online Banking	*Coded as 1 if customer uses Internet banking facilities*
• Credit Card	*Coded as 1 if customer uses credit card issued by Universal Bank*

Among these 5000 customers, only 480 (= 9.6%) accepted the personal loan offered to them in a previous campaign. The goal is to build a model that identifies customers who are most likely to accept the loan offer in future mailings.

The bank application scenario is just a selected example; other application scenarios is in a sales campaign for example, where the goal is to build a model that identifies customers who are most likely to make a purchase. In a service campaign, that is, a business that offers one form of service or another to clients, the goal is to build a model that identifies clients who will take the service offer in future proposals or marketing.

The outcome variable is personal loan, with *yes* defined as the *success* (this is equivalent to setting the outcome variable to 1 for an acceptor and 0 for a non-acceptor).

Some data preprocessing necessary for solving this problem is to convert the attribute *Education* to polynomial first and then to dummy later and remove one of the dummies. The target attribute *Personal loan* will be converted to binomial. After this, the data should be visualized for classification and use the correlation matrix to discover correlated attributes (this will be skipped in this practical demonstration as it is captured in exercise 4 of this chapter). Also, as in linear regression, in logistic regression we can use automated variable selection heuristics such as stepwise selection, forward selection, and backward elimination (more explanations in [4,5,6]).

1. Create a new process. **Import** the data named *Bank.xlsx*. Exclude attributes *ID column* and *ZIP Code*. Convert *Education* to polynominal. Convert *Personal loan* to binominal.

2. Create a new process, and drag the already imported *Bank.xlsx* data to the design view. Connect the data to *res*, run the process, and view the **Statistics** (notice that there are no missing values and all conversions done in step 1 has been taken care of). We will use the settings in Figures 7-23a (main process) and 7-23b (inner process) to develop and evaluate the model.

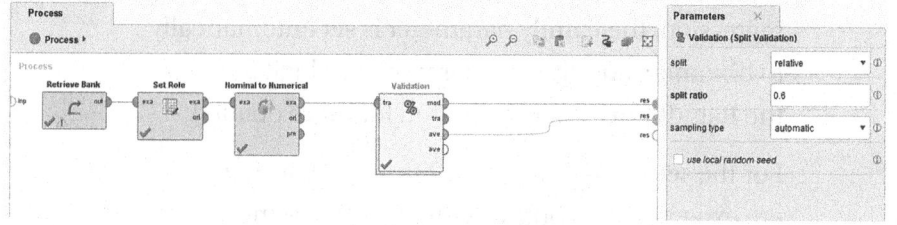

Figure 7-23a. *Main process (logistic regression)*

In Figure 7-23a, click the **Set Role** operator and set the *attribute name* to be *Personal loan* and *target role* to be label. For the **Nominal to Numerical** operator, you will select the attribute *Education* (convert to dummies). For the **Split Validation** operator, set the split ratio to 0.6.

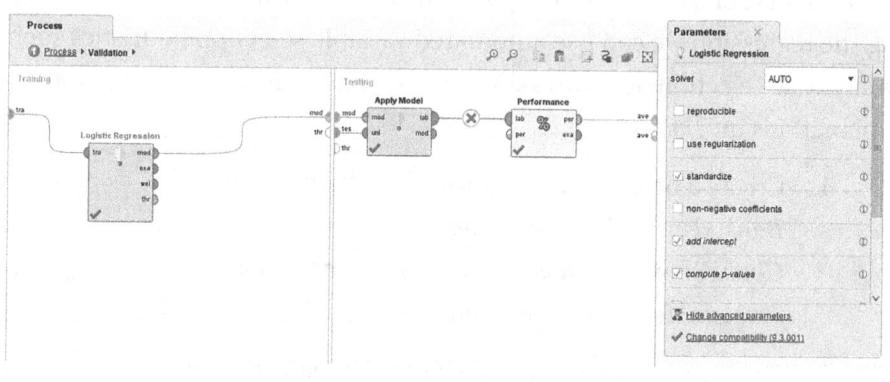

Figure 7-23b. *Inner process (logistic regression)*

In Figure 7-23b for the **Logistic Regression** operator, the **Apply Model** operator, and the **Performance (Classification)** operator, we will leave as default settings. The **Logistic Regression** operator is a simplified version of the Generalized Linear Model operator.[7] To perform logistic regression, the Family parameter is set automatically to binomial, and the link parameter to logit (check the RapidMiner documentation for more details).

For the settings in Figures 7-23a and 7-23b, we are expecting two outputs. Figure 7-24a is the logistic regression model, and Figure 7-24b is the performance of the model.

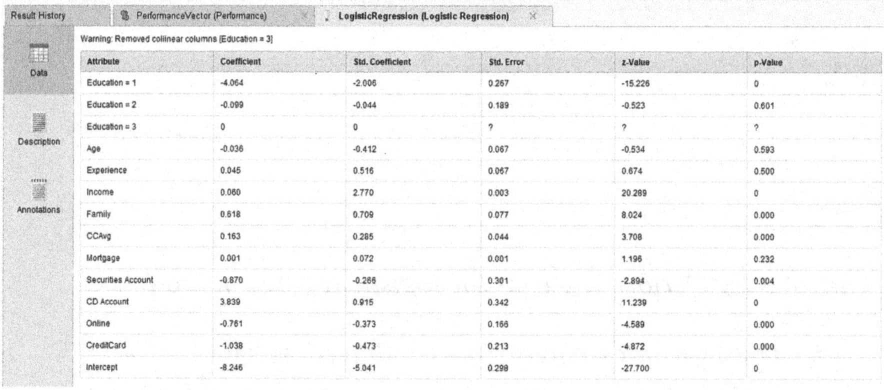

Attribute	Coefficient	Std. Coefficient	Std. Error	z-Value	p-Value
Education = 1	-4.064	-2.006	0.267	-15.226	0
Education = 2	-0.099	-0.044	0.189	-0.523	0.601
Education = 3	0	0	?	?	?
Age	-0.038	-0.412	0.067	-0.534	0.593
Experience	0.045	0.516	0.067	0.674	0.500
Income	0.060	2.770	0.003	20.289	0
Family	0.618	0.709	0.077	8.024	0.000
CCAvg	0.163	0.285	0.044	3.708	0.000
Mortgage	0.001	0.072	0.001	1.196	0.232
Securities Account	-0.870	-0.266	0.301	-2.894	0.004
CD Account	3.839	0.915	0.342	11.239	0
Online	-0.761	-0.373	0.166	-4.589	0.000
CreditCard	-1.038	-0.473	0.213	-4.872	0.000
Intercept	-8.246	-5.041	0.298	-27.700	0

Warning: Removed collinear columns [Education = 3]

Result History · PerformanceVector (Performance) · LogisticRegression (Logistic Regression)

Figure 7-24a. *Logistic regression model*

3. From Figure 7-24a, note that some of these variables might be removed based on their p value; very small p values are usually recommended (0.001 or less).[8] Also, Education=3 has been removed due to multicollinearity. The positive coefficients for the variables CD count and mortgage mean that having cd account and mortgage are associated with higher probabilities of accepting the loan offer (they are dummy variables indicated by 1). In contrast, having securities, online, creditCard, etc. is associated with lower acceptance rates. For the continuous predictors, positive coefficients indicate that a higher value on that predictor is associated with a higher probability of accepting the loan offer (e.g., income: higher-income customers tend more to accept the offer). Similarly, negative coefficients indicate that a higher value on that predictor is associated with a lower probability of accepting the loan offer (e.g., age).

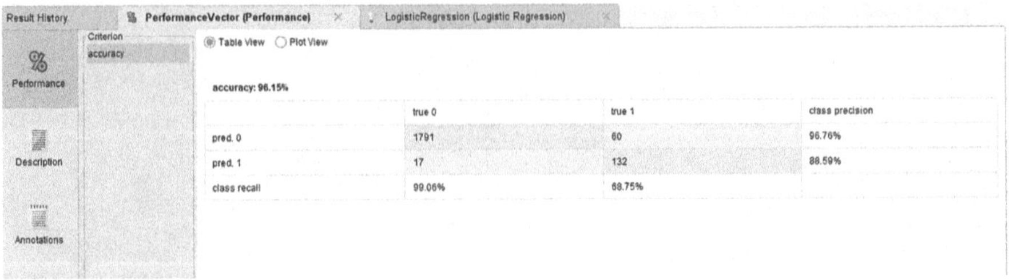

Figure 7-24b. *Performance for the logistic regression model*

4. Once we have established an optimal model for
    prediction, we can now use the model to predict for
    new data (newBank.xlsx).

Customer profiling using logistic regression

Logistic models (e.g., Figure 7-24a) can be used to provide useful
information about the roles played by different attributes in classification.
If we want to know how increasing family income by one unit will affect
the probability of loan acceptance, we can use Equation 7.14, modified for
a single predictor or attribute as seen in Equation 7.15.

$$Odds = e^{\beta_0 + \beta_1 x_1 + \beta_2 x_2 + \dots + \beta_q x_q} \qquad \text{(Eq. 7.14)}$$

$$Odds\left(Income = x\right) = e^{\beta_0 + \beta_1 X} \qquad \text{(Eq. 7.15)}$$

The estimated coefficients for the single predictor model are
β_0 = -6.127 and β_1 = 0.037 as revealed in the model in Figure 7-25

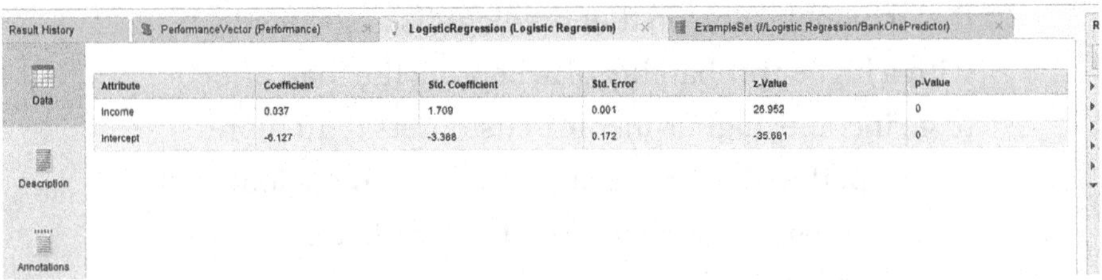

Figure 7-25. *Logistic regression model for only variable (income)*

The odds that a customer with income zero will accept the loan are estimated by $e^{-6.127+(0.037)(0)} = 0.00218$.

The odds of accepting the loan with an income of \$100K are $e^{-6.127+(0.037)(100)} = 0.0883$.

7.8 Problems

1. Write the logistic regression equation from the model in Figure 7-24a.

2. Complete the classification example using logistic regression (Section 7.7) by predicting for new customers.

3. Repeat the practical business problem III; this time around, remove attributes NumOfProducts and EstimatedSalary and compare the results. Any improvement in the performance of the algorithms?

4. Repeat the practical business problem III; this time around, include the correlation matrix and use it to remove correlated attributes before applying all three algorithms. Is the accuracy increased?

5. For this exercise, we will use a case study of a data science for a business training outfit called iandfnetworksolutions. iandfnetworksolutions is an online data analytics training and consulting with less than ten members of staff. The task at hand for the data scientist is to come up with the best subset of prospects (10%) that they should target for a one-on-one campaign, which they have found out to be the most effective in this situation. Internal data is named InternalData.xlsx; external data is Prospects.xlsx.

6. For this exercise, we will use a case study of a large noodle retail store in Nigeria. They are a large store that sells different types of noodles to smaller stores that retail in more smaller quantity. The retail store has been able to keep the records of their customers scattered all over Nigeria and has been able to come up with the data of likely prospects. The task at hand for the data scientist is to come up with the best model to

   a. Classify prospects based on whether to target them for online sales or offline sales (classification)

   b. Find factors that differentiate between online and offline customers (profiling). Data files are named

      - *Internal data*: NoodleRetailLRInternal.xlsx

      - *External data*: NoodleRetailLRExternal.xlsx

7.9 References

1. Jiawei Han, Micheline Kamber and Jian Pei (2012) Data Mining Concepts and Techniques, Third Edition, published by Elsevier.

2. NeuralMarketTrends (May 25, 2018) Balancing Data in RapidMiner, `www.youtube.com/watch?v=fpfvNhWmmQo`

3. Galit Shmueli, Nitin R. Patel, & Peter C. Bruce, Data Mining for Business Analytics, Concepts, Techniques and Applications in R, published by John Wiley & Sons, Inc., Hoboken, New Jersey, 2018, `www.stat.berkeley.edu/users/breiman/RandomForests/cc_home.htm`

4. Understand Forward and Backward Stepwise Regression, George Choueiry, https:// quantifyinghealth.com/stepwise-selection/

5. Selection Process for Multiple Regression, www. statisticssolutions.com/free-resources/ directory-of-statistical-analyses/selection- process-for-multiple-regression/

6. www.researchgate.net/publication/331749857_ Generalized_Linear_Models

7. P-Value in Regression by Priya Pedamkar, www. educba.com/p-value-in-regression/

8. Galit Shmueli, Nitin R. Patel, & Peter C. Bruce, Data Mining for Business Analytics, Concepts, Techniques and Applications in R, published by John Wiley & Sons, Inc., Hoboken, New Jersey, 2018.

More resources on the chapter for further reading

- Target Customers For Business Success (www.youtube. com/watch?v=9XJybFFGvsg&t=1794s)

- Creating Lift Charts using RapidMiner Studio (www. youtube.com/watch?v=zi8urIdAb5A)

- A Gentle Introduction to the Fbeta-Measure for Machine Learning by Jason Brownlee on February 24, 2020, https://machinelearningmastery.com/fbeta- measure-for-machine-learning/

- Boxplots vs. Individual Value Plots: Comparing Groups by Jim Frost, https://statisticsbyjim.com/basics/ graph-groups-boxplots-individual-values/

CHAPTER 8

Advanced Descriptive Analytics

This chapter is focused mainly on advanced descriptive analytics techniques. In this chapter, we will first explain the concept of clustering which is a type of unsupervised learning approach. We will then pick one clustering technique which is the k-means clustering. Using the fourth practical business problem, we will explain how we can use the k-means clustering technique to solve a real business problem. Next, we will explain the association rule example and finally network analysis. We will focus the explanation of these techniques on solving business-related problems, particularly for small businesses, and conclude with the fifth business problem which is focused on using network analytics for employee efficiency.

8.1 Clustering

Clustering is an unsupervised machine learning approach that aims to segment data into similar groups of records to gain insight. Cluster analysis is a method that forms clusters of records by calculating the distance between these records. Clustering is applied in business[2] in various ways which include categorizing clients based on demographic and transaction history information and then tailoring a marketing plan for each category. It can also be used to create segments for customers and give them

© Afolabi Ibukun Tolulope 2022
A. I. Tolulope, *Data Science and Analytics for SMEs*,
https://doi.org/10.1007/978-1-4842-8670-8_8

identities called "personas" to help focus energy on sales. It is also used in market structure analysis to find groups of comparable products based on competitive similarity measurements.

There are two popular clustering approaches:[4]

1. *Hierarchical method*: This has to do with arranging clusters into a natural hierarchy; hierarchical approaches are extremely useful. These methods are agglomerative or divisive in nature.

2. *Nonhierarchical method*: The k-means algorithm is an example of such a technique. K signifies that the algorithm assigns records to each cluster based on a predetermined number of clusters. Because it is less computationally costly, thus it's better for big data.[3]

Regardless of whether the clustering approach is hierarchical or nonhierarchical, two sorts of distances must be defined: distance between records and distance between two clusters in the dataset used. In addition, there is a range of metrics that can be employed for these distances in both methods.

It is necessary to keep in mind the following while measuring the distance between two records:

- The distance should not be negative.

- There should be no distance between a record and itself (0).

- The distance between i and j must be equal to the distance between j and i.

- Any pair's distance cannot exceed the total of the distances of the other two pairings.

The measures used for the distance between two records (numerical data) include the following:

- *Euclidean distance*: This type of measure requires the normalization of continuous measurements before computing. A different distance (such as the statistical distance) is also recommended if the measurements are highly correlated. Outliers can make Euclidean distance skewed. For Euclidean distance, outliers are either deleted or more robust distances like Manhattan distances are used.

- Correlation-based similarity.

- Statistical distance which is also referred to as Mahalanobis distance.

- Manhattan distance ("city block").

- Maximum coordinate distance.

For more information on the preceding list, check Han et al. (2012).

Similarity measures are more intuitively appealing than distance measures for measuring the distance between two records having categorical data with binary values. In this case, the matching coefficient and Jaccard's coefficient are the most helpful similarity measurements. For measuring the distance between two records, categorical data and mixed data (some continuous and some binary), a similarity coefficient suggested by Gower is very useful.[5]

When measuring the distance between two clusters, the following definitions are important:

- *Minimum distance*: The distance between the two clusters' closest pair of records, one record in each cluster.

- *Maximum distance*: The distance between the two records that are the farthest apart.

- *Average distance*: The average distance between any two records.

- *Centroid distance*: Centroid linkage clustering is based on centroid distance, in which groups are represented by their mean values for each variable, forming a vector of means. The distance between these two vectors determines the distance between two clusters.

It is good for clusters to be well separated; the more separated, the more homogenous groups we have, so these distances help us to see this separation clearly.

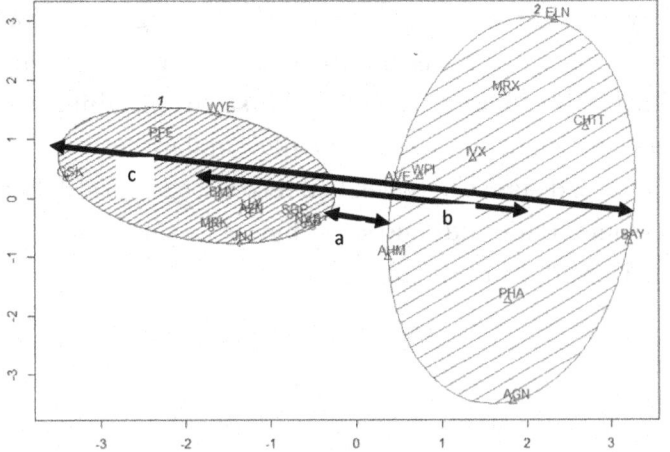

Figure 8-1. *A two-dimensional illustration of various distance metrics between two clusters: (a) minimum distance, (b) centroid distance, and (c) maximum distance*

8.2 K-Means

The k-means algorithm divides the sample into a predefined number k of well-separated (no overlaps) clusters with the purpose of making these clusters similar in terms of the measures adopted (e.g., Figure 8-1). The total distance (or sum of squared Euclidean distances) between records and their cluster centroid is a commonly used metric of within-cluster dispersion. The k-means algorithm is expressed in Figure 8-2.

1.To begin, create k initial clusters (user chooses k).
2. Each record is moved to the cluster with the "nearest" centroid at each stage.
3. Recalculate the centroids of clusters that have acquired or lost a record, and repeat Step 2.
4. Stop when moving any more records between clusters increases cluster dispersion.

Figure 8-2. K-means clustering algorithm

How to determine k

1. K can be determined by external considerations such as previous or domain knowledge and some constraints presented in the business environment.

2. You can try different k and compare the resulting distance. The distance with cluster members has to be small, and the distance between different clusters has to be large.

3. K can also be determined by direct methods that optimize a criterion, such as the within-cluster sums of squares or the average silhouette. These methods include *gap statistic* methods; there are more than 30 other indices. Figure 8-3 demonstrates using three of these methods after which we take a

vote. Based on Figure 8-3, the answer is between 4 (elbow) and 2 (silhouette); to make your decision, you have to try several other indices, and the number of clusters with the highest vote wins.

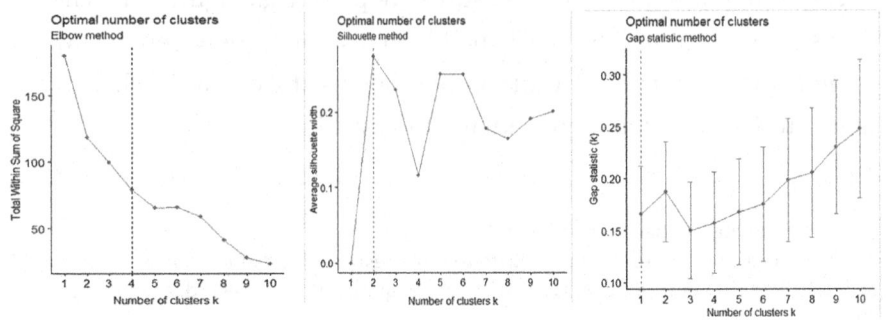

Figure 8-3. *Comparing the results of three direct methods to determine the best k*

Validating clusters

Finding meaningful clusters is a key goal of cluster analysis. The following steps can assist you in accomplishing this:

1. *Cluster interpretability*: Is the interpretation of the clusters that resulted reasonable? Explore the properties of each cluster to interpret the clusters. You can do this by doing the following:

   a. Analyze the results of the cluster analysis to derive summary statistics (such as averages, minimums, and maximums) for each cluster's measurement.

   b. Examine the clusters for separation along with some common feature (variable) that was omitted from the cluster analysis.

    c. Label the clusters. In this step, we're going to name and label the different clusters depending on our understanding of the data.

2. *Stability of the cluster*: Are cluster assignments affected greatly if some of the inputs are changed?

3. *Cluster separation*: Determine whether the separation is fair by examining the ratio of between-cluster variation to within-cluster variation. This is where we need the distance between clusters. Internal validity measurements are concerned with the information included in the clusters and the arrangement of data points with regard to them. It is important that all the points in a cluster are close to one another in order to have good clustering. It is also important that clusters are well separated and distinct from one another. The validity measures used in internal validity include the following:

- Average within centroid distance

- The density of clusters

- Distribution of items by sum of squares

- Davies-Bouldin index

4. *Number of clusters*: The number of clusters generated must be relevant to the goal of the analysis. External validity measures are used here to compare the clusters created by the clustering algorithm with previously known clusters. To do this in real-world scenarios, model development and

testing require input from experts who will have to provide the clusters. The measures used for external validity include

- Rand index[7]

- Adjusted Rand index[6]

- Jaccard index[8]

- Fowlkes-Mallows index[8]

5. There is also a density-based approach (e.g., similarity measure).

For more on clusters, please check Han et al. (2012).

Interpreting the clusters

- We look at the cluster centroids to determine the characteristics of the generated clusters.

- In addition, we can see which variables are most effective at distinguishing the clusters.

- We can also study the information about the within-cluster dispersion.

- Using the Euclidean distance between the clusters' centroids, we can determine how far apart the clusters are.

Interpreting the clusters for target marketing purpose

As soon as we have a clear picture of the numerous client personas, we can tailor our marketing interactions to each persona's unique preferences, that is, linking the newly found client personas to the most appropriate marketing interactions for each of them. These interactions should be tailored to meet the unique requirements, wants, and preferences of each individual persona.

8.3 Practical Business Problem IV (Customer Segmentation)

The case study of this practical business problem is a noodle retail small business. NoodleRetail.xlsx is a data on a noodle retail company that intends to segment their current customers so as to be able to describe them. The goal is to segment and describe current customers, that is, identify clusters of customers that have similar characteristics for the purpose of targeting different segments for different types of offers in the future. To do this, the clustering process needs to give them segments that are as homogeneous as possible within themselves and heterogeneous from each other.

Customer segmentation using k-means

The following steps are used for the segmentation process:

1. *Data preprocessing*: We will first handle missing values if there are any as k-means cannot deal with missing values. If there are only few missing values, they can be excluded; if there are many, they have to be imputed. We will then perform data visualization to discover outliers. We will convert categorical to numerical (as k-means uses a distance function and cannot handle categorical variables directly). If the categorical variables are ordinal, we will replace with an arithmetic sequence, and if it's nominal, convert to binary. After this, we will normalize the data and apply an appropriate dimension reduction technique. It is important to reduce the dimension because any more than few tens of dimensions mean that distance interpretation isn't obvious and must be prevented. For this example, we use Section 2.5 as a guide for the data preprocessing for unsupervised learning.

2. Determine the number of k. There are several approaches to doing this (as taught earlier), but we shall be supplying the number of k to use for this assignment as a domain expert.

3. Perform k-means clustering using an appropriate distance measure for the distance between records (numerical measure was selected because we have converted all the attributes to numerical).

4. Validate the clusters. To validate the generated cluster, we use the *avg. within centroid distance*. (To further check the stability of the clusters, you can remove a random 5% of the data (by taking a random sample of 95% of the records) and repeat the analysis. Does the same picture emerge?)

5. To interpret the clusters, we will

   - Compare the cluster centroid to characterize the different clusters, and try to give each cluster a label.

   - Provide an appropriate name for each cluster using any or all of the variables in the dataset.

   - Which clusters would you target for offers, and what types of offers would you target to customers in that cluster?

Data description

The noodle retail data consist of the following attributes:

Brand

Country

Customer Id

Gender

Income Class

List Price

Number of Records

Online Order

Owns Car

Product Class

Product Id

Product Size

Purchases

Past 3 Years

State

Transaction Date

Transaction Id

Deceased Indicator

Tenure

1. (a) *Data preprocessing*: Create a repository (**Clustering**); create two subfolders in the repository and name them data and process. **Import** the data named NoodleRetail.xlsx into the **data** subfolder. When importing, make sure to exclude the following attributes based on the reasons stated:

    - Country (same values all through)

    - Number of Records (same values all through)

    - Customer Id (not needed for the preprocessing)

    - Transaction Date (not needed for clustering)

    - Transaction Id (not needed for clustering)

    - Deceased Indicator (same values all through)

    - Product Id (not needed for the preprocessing or clustering)

Figure 8-4 shows the **Statistics** of the data.

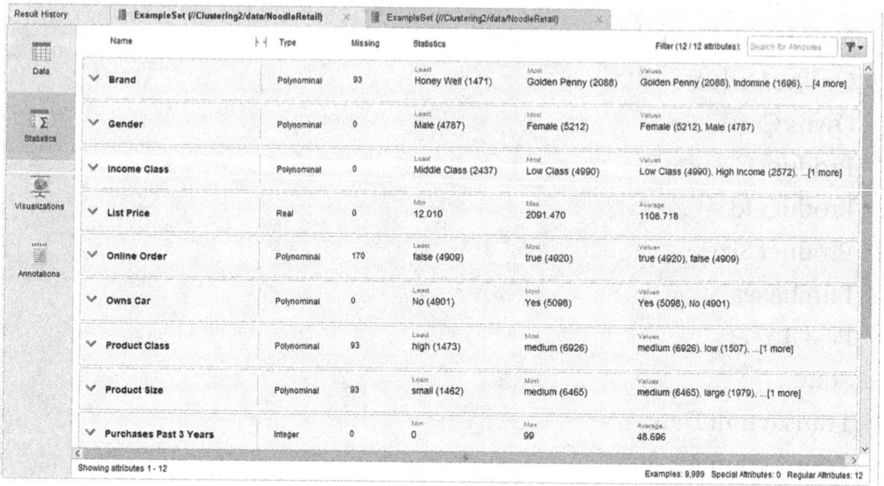

Figure 8-4. *Statistics of the data*

The following missing values are obvious from these statistics and will be treated by replacing them with the average value of the attributes:

- *Brand-93*

- *Online order-170*

- *Product class-93*

- *Product size-93*

- *Tenure-214*

Create a new process (call it *preprocessing1*). Use the settings in Figure 8-5 to replace the missing value with average. Click the **Replace Missing Values** operator and select Brand, Online order, Product class, Product size, and Tenure.

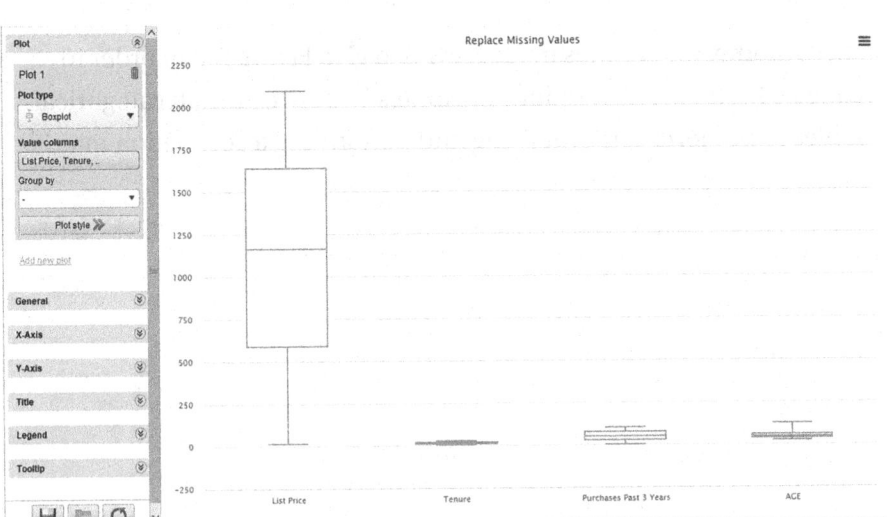

Figure 8-5. *Process to deal with missing values*

Run the process and visualize the resulting data (only the numerical attributes) for outliers using the settings in Figure 8-6.

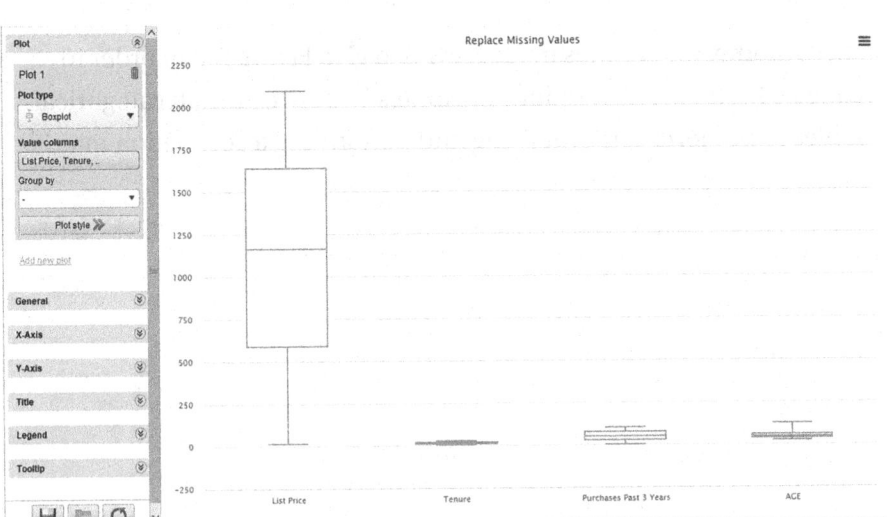

Figure 8-6. *Visualizing the numerical attributes*

Whether the numerical variables are visualized separately or together as revealed in the settings in Figure 8-6, there appear to be no outliers.

Figure 8-7 is the visualization of the age attribute using a scatter plot to detect outliers; this can be done for all the numerical attributes.

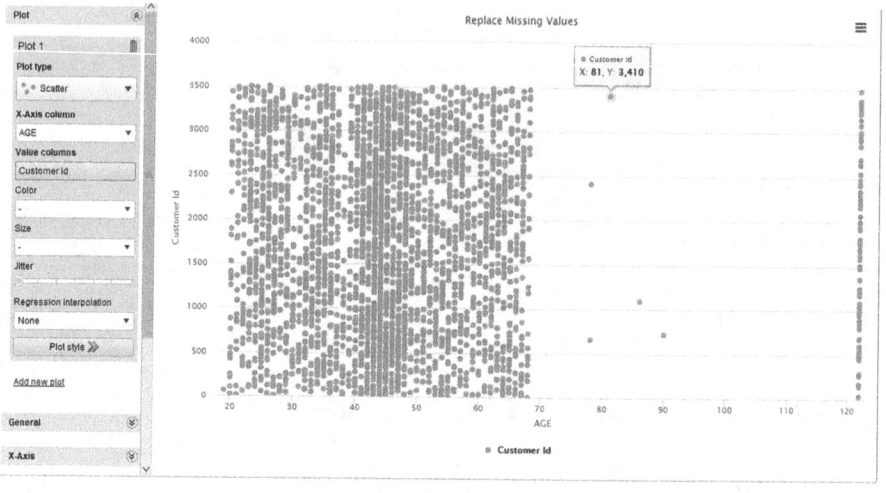

Figure 8-7. *Scatter plot of the numerical attribute age*

Let's modify the settings in Figure 8-5 to give Figure 8-8 in order to investigate for the need to reduce attributes by converting all categorical variables to numerical, normalizing and performing a correlation matrix.

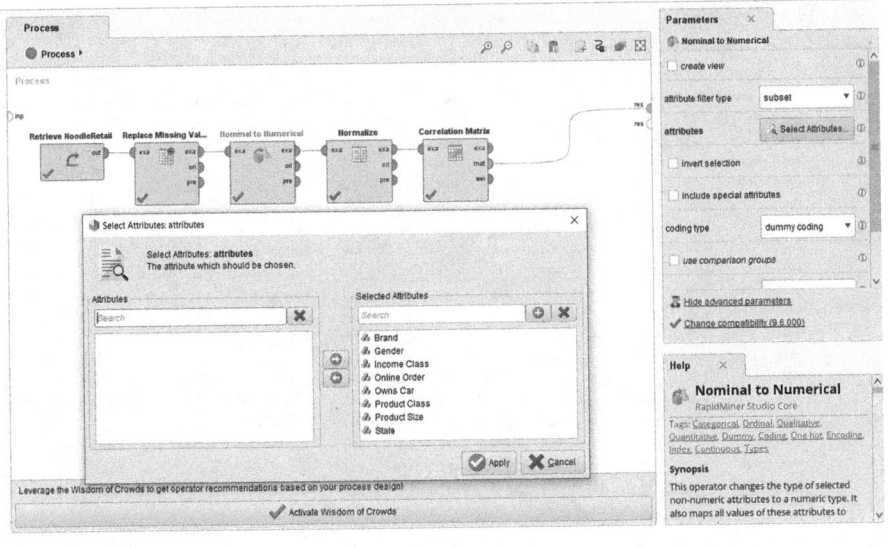

Figure 8-8. *Further preprocessing of the data*

As seen in Figure 8-8, the Nominal to Numerical operator selects all the categorical attributes. Click the **Normalize** operator, and for the **attribute filter type** parameter, select **all**. Leave the **Correlation Matrix** operator as the default setting. Running the process gives us the correlation matrix in Figure 8-9; clicking the **Pairwise Table** in the same figure and sorting the correlation values in descending order, we get Figure 8-10.

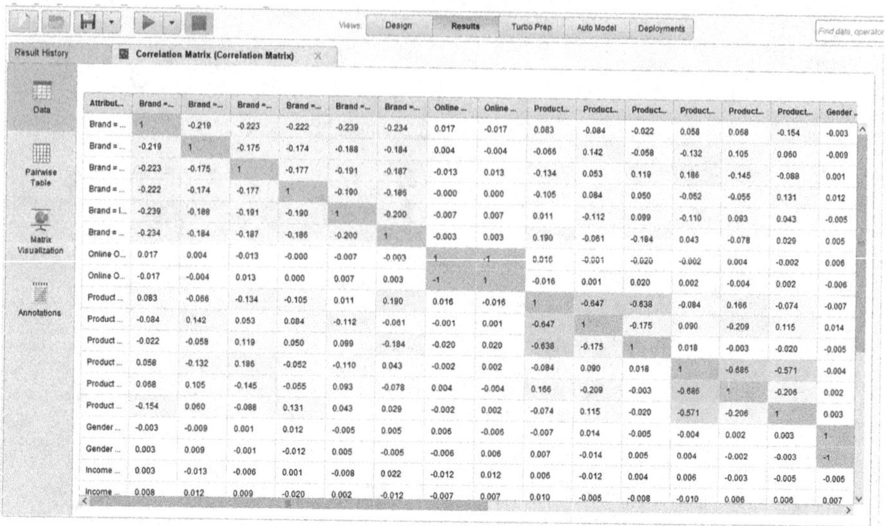

Attribut...	Brand =...	Brand =...	Brand =...	Brand =...	Brand =...	Brand =...	Online ...	Online ...	Product...	Product...	Product...	Product...	Product...	Product...	Gender ...
Brand = ...	1	-0.219	-0.223	-0.222	-0.239	-0.234	0.017	-0.017	0.083	-0.084	-0.022	0.058	0.068	-0.154	-0.003
Brand = ...	-0.219	1	-0.175	-0.174	-0.188	-0.184	0.004	-0.004	-0.066	0.142	-0.058	-0.132	0.105	0.060	-0.009
Brand = ...	-0.223	-0.175	1	-0.177	-0.191	-0.187	-0.013	0.013	-0.134	0.053	0.119	0.186	-0.145	-0.088	0.001
Brand = ...	-0.222	-0.174	-0.177	1	-0.190	-0.186	-0.000	0.000	-0.105	0.084	0.050	-0.052	-0.055	0.131	0.012
Brand = I...	-0.239	-0.188	-0.191	-0.190	1	-0.200	-0.007	0.007	0.011	-0.112	0.099	-0.110	0.093	0.043	-0.005
Brand = ...	-0.234	-0.184	-0.187	-0.186	-0.200	1	-0.003	0.003	0.190	-0.061	-0.184	0.043	-0.078	0.029	0.005
Online O...	0.017	0.004	-0.013	-0.000	-0.007	-0.003	1	-1	0.016	-0.001	-0.020	-0.002	0.004	-0.002	0.006
Online O...	-0.017	-0.004	0.013	0.000	0.007	0.003	-1	1	-0.016	0.001	0.020	0.002	-0.004	0.002	-0.006
Product ...	0.083	-0.066	-0.134	-0.105	0.011	0.190	0.016	-0.016	1	-0.647	-0.638	-0.084	0.166	-0.074	-0.007
Product ...	-0.084	0.142	0.053	0.084	-0.112	-0.061	-0.001	0.001	-0.647	1	-0.175	0.090	-0.209	0.115	0.014
Product ...	-0.022	-0.058	0.119	0.050	0.099	-0.184	-0.020	0.020	-0.638	-0.175	1	0.018	-0.003	-0.020	-0.005
Product ...	0.058	-0.132	0.186	-0.052	-0.110	0.043	-0.002	0.002	-0.084	0.090	0.018	1	-0.686	-0.571	-0.004
Product ...	0.068	0.105	-0.145	-0.055	0.093	-0.078	0.004	-0.004	0.166	-0.209	-0.003	-0.686	1	-0.206	0.002
Product ...	-0.154	0.060	-0.088	0.131	0.043	0.029	-0.002	0.002	-0.074	0.115	-0.020	-0.571	-0.206	1	0.003
Gender ...	-0.003	-0.009	0.001	0.012	-0.005	0.005	0.006	-0.006	-0.007	0.014	-0.005	-0.004	0.002	0.003	1
Gender ...	0.003	0.009	-0.001	-0.012	0.005	-0.005	-0.006	0.006	0.007	-0.014	0.005	0.004	-0.002	-0.003	-1
Income ...	0.003	-0.013	-0.006	0.001	-0.008	0.022	-0.012	0.012	0.006	-0.012	0.004	0.006	-0.003	-0.005	-0.005
Income ...	0.008	0.012	0.009	-0.020	0.002	-0.012	-0.007	0.007	0.010	-0.005	-0.008	-0.010	0.006	0.006	0.007

Figure 8-9. *Correlation matrix*

Result History		Correlation Matrix (Correlation Matrix)	

Data

Pairwise Table

Matrix Visualization

Annotations

First Att...	Second ...	Corr... ↓
Tenure	AGE	0.331
Product ...	List Price	0.309
Brand = ...	Product ...	0.190
Brand = ...	Product ...	0.186
Product ...	Product ...	0.166
Brand = ...	Product ...	0.142
Brand = ...	Product ...	0.131
Brand = ...	List Price	0.122
Product ...	List Price	0.120
Brand = ...	Product ...	0.119
Product ...	Product ...	0.115
Brand = ...	Product ...	0.105
Product ...	List Price	0.105
Gender ...	AGE	0.101
Brand = I...	Product ...	0.099
Brand = I...	Product ...	0.093
Brand = I...	List Price	0.092
Product ...	Product ...	0.090

Figure 8-10. *Correlation matrix (Pairwise Table)*

From Figure 8-10, we see that there are no strongly correlated pair (the highest pairwise correlation is 0.331), and for this reason, all the attributes will be used for this demonstration. (Some attributes on top of the list might be removed later to see if there will be better cluster separation when evaluated.) This decision is concluded based on the Pearson correlation formula, which is used by the ***Correlation Matrix*** operator.

Pearson's correlation coefficient R is the end result obtained in which the value must range from –1 to +1 or –1<=R<=1.

$$r = \frac{n\Sigma xy - \Sigma x \Sigma y}{\sqrt{\left\{n\Sigma x^2 - \left(x\Sigma^2\right)\right\}\left\{n\Sigma y^2 \left(\Sigma y\right)^2\right\}}}$$

The preceding formula is used to get the value for R, and the value for R must fall in the following range:

- *+1 to 0.7*: Means that the strength of the variables is perfectly related in a positive way

- *0.6 to 0.5*: Means that the strength of variables is fairly related

- *0.4 to 0.2*: Means that the strength of variables is poorly related

- *-1 to -0.7*: Means that the strength of variables is strongly related in a negative way

- *-0.6 to -0.5*: Means that the strength of variables is fairly related

- *-0.4 to -0.2*: Means that the strength of variables is poorly related

- *0*: Means that there is no strength or relation

(b) We then move to the second part of preprocessing which deals with importing the afresh (because we need the customer ID for clustering) and selecting only the attributes that are not strongly correlated. The result of this process will be stored in a csv file which will then be imported again for k-means clustering.

Import the data (NoodleRetail.xlsx) again; this time, change the role of customer Id to Id as seen in Figure 8-11. After this, exclude the attributes Country, Number of Records, Transaction Date, Transaction Id, Deceased Indicator, and Product Id. Name this new data NoodleRetail2 in RapidMiner.

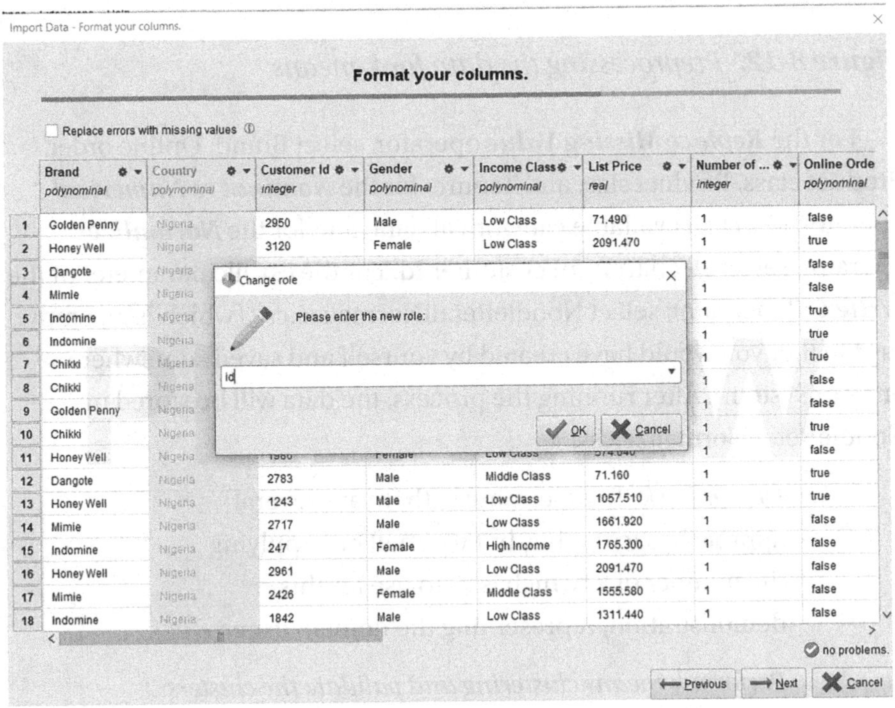

Figure 8-11. *Importing the data for k-means clustering*

Create a new process (call it preprocessing2). Use the settings in Figure 8-12 to create the process. The purpose of this process is to preprocess only the attributes that will be clustered.

217

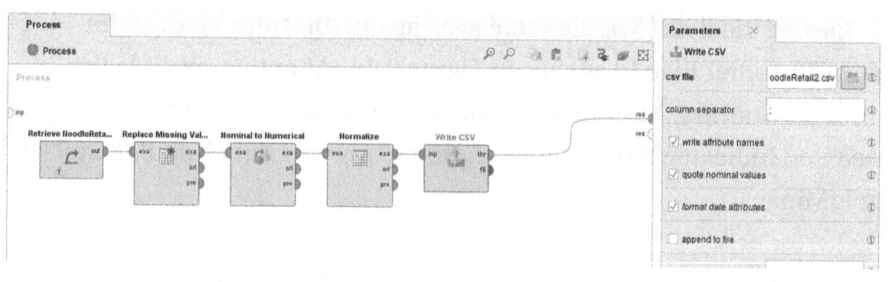

Figure 8-12. *Preprocessing the data for k-means*

For the **Replace Missing Value** operator, select Brand, Online order, Product class, Product size, and Tenure; for the **Nominal to Numerical** operator, select all available categorical data; and for the **Normalize** operator, select all apart from customer Id. For the csv file parameter of the **Write CSV** operator, select NoodleRetailNormalize.csv (which is an empty csv file that you would have created by yourself and saved somewhere on your system). After running the process, the data will be stored in NoodleRetailNormalize.csv.

1. *Determine k*: To determine k, there are several approaches to do this, but we shall be supplying the number of k (which is 2) to use for this demonstration, representing the domain expert.

2. *Perform k-means clustering and validate the clusters*: Numerical measures were selected because we have converted all the attributes to numerical. **Import** the NoodleRetailNormalize.csv, and create a new process (call it kmeans); use the settings in Figure 8-13 for this process (the k-means clustering). For the k parameter of the **k means** operator, set it as 2 and leave all other parameter settings as default.

For the ***Cluster Distance Performance*** operator, leave the main criterion parameter as default. (Please read the help of these operators in the RapidMiner documentation for more details.)

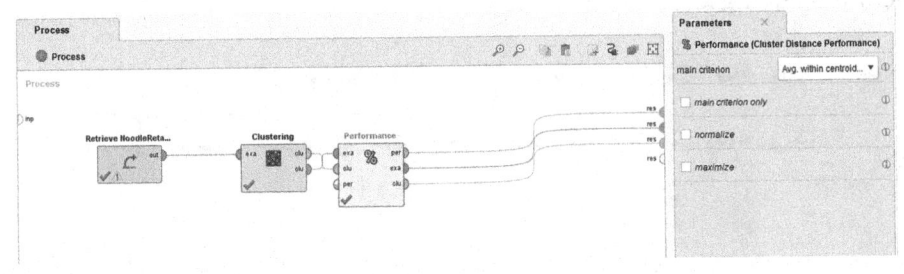

Figure 8-13. *k-means process*

From Figure 8-13, we expect three outputs as shown in Figures 8-14a, 8-14b, and 8-14c, respectively.

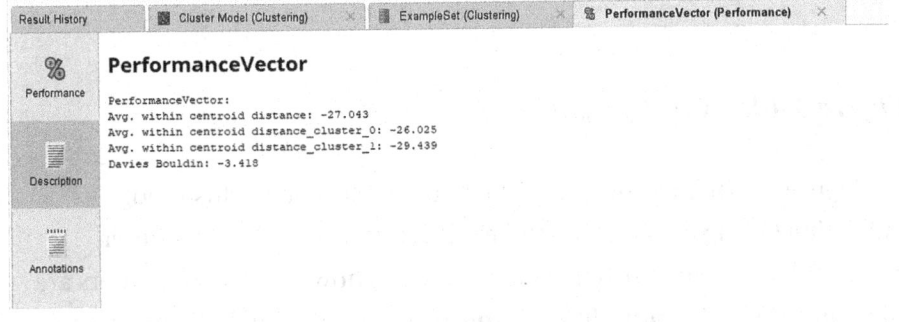

Figure 8-14a. *Performance results for the k-means model*

Figure 8-14a is the performance of the k-means model which is used to know how well the clusters are separated. From the RapidMiner documentation, avg._within_centroid_distance, the average within-cluster distance is calculated by averaging the distance between the centroid and all examples of a cluster. The smaller, the better. When comparing the results of several k-means models, the model that produces a collection of clusters with the smallest Davies-Bouldin index is considered the best algorithm based. Even though we chose a value for our k in this demonstration, it is advised in practice to try several values of k and compare the avg._within_centroid_distance and the Davies-Bouldin index before selecting k.

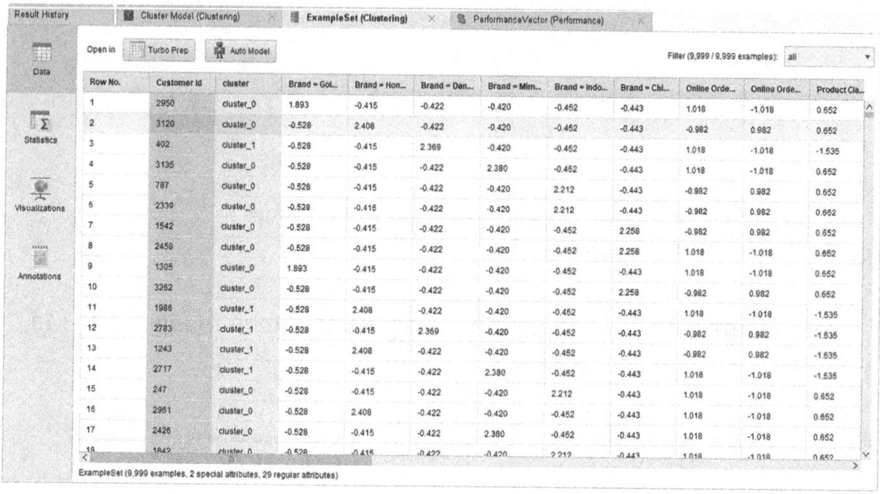

Row No.	Customer Id	cluster	Brand = Gol...	Brand = Hon...	Brand = Dan...	Brand = Mim...	Brand = Indo...	Brand = Chi...	Online Orde...	Online Orde...	Product Cla...
1	2950	cluster_0	1.893	-0.415	-0.422	-0.420	-0.452	-0.443	1.018	-1.018	0.652
2	3120	cluster_0	-0.528	2.408	-0.422	-0.420	-0.452	-0.443	-0.982	0.982	0.652
3	402	cluster_1	-0.528	-0.415	2.369	-0.420	-0.452	-0.443	1.018	-1.018	-1.535
4	3135	cluster_0	-0.528	-0.415	-0.422	2.380	-0.452	-0.443	1.018	-1.018	0.652
5	787	cluster_0	-0.528	-0.415	-0.422	-0.420	2.212	-0.443	-0.982	0.982	0.652
6	2339	cluster_0	-0.528	-0.415	-0.422	-0.420	2.212	-0.443	-0.982	0.982	0.652
7	1542	cluster_0	-0.528	-0.415	-0.422	-0.420	-0.452	2.258	-0.982	0.982	0.652
8	2459	cluster_0	-0.528	-0.415	-0.422	-0.420	-0.452	2.258	1.018	-1.018	0.652
9	1305	cluster_0	1.893	-0.415	-0.422	-0.420	-0.452	-0.443	1.018	-1.018	0.652
10	3262	cluster_0	-0.528	-0.415	-0.422	-0.420	-0.452	2.258	-0.982	0.982	0.652
11	1986	cluster_1	-0.528	2.408	-0.422	-0.420	-0.452	-0.443	1.018	-1.018	-1.535
12	2783	cluster_1	-0.528	-0.415	2.369	-0.420	-0.452	-0.443	-0.982	0.982	-1.535
13	1243	cluster_1	-0.528	2.408	-0.422	-0.420	-0.452	-0.443	-0.982	0.982	-1.535
14	2717	cluster_1	-0.528	-0.415	-0.422	2.380	-0.452	-0.443	1.018	-1.018	-1.535
15	247	cluster_0	-0.528	-0.415	-0.422	-0.420	2.212	-0.443	1.018	-1.018	0.652
16	2961	cluster_0	-0.528	2.408	-0.422	-0.420	-0.452	-0.443	1.018	-1.018	0.652
17	2426	cluster_0	-0.528	-0.415	-0.422	2.380	-0.452	-0.443	1.018	-1.018	0.652

Figure 8-14b. *Cluster dataset*

Figure 8-14b is the output of the data that has been clustered, indicating (using the cluster Id) the cluster to which each customer belongs. This can be further visualized to see how well the customers are separated. (The ***Cluster Model Visualizer*** operator can help, but it does not come with the free edition of RapidMiner.)

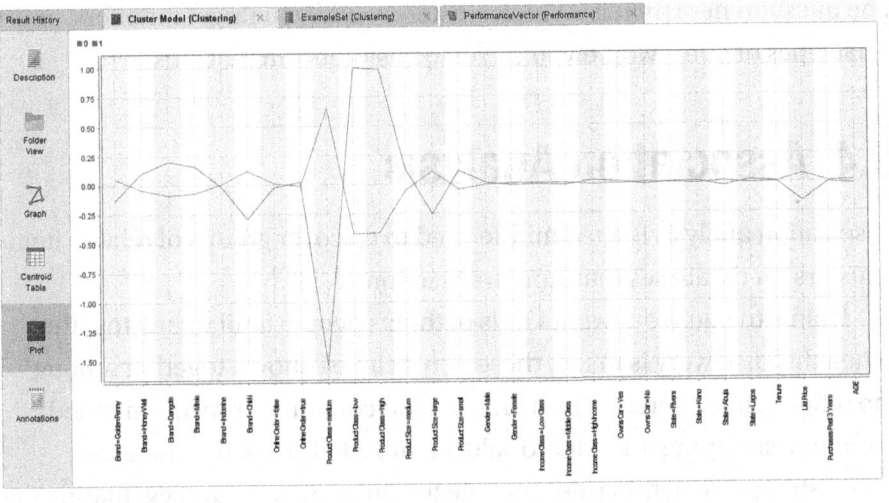

Figure 8-14c. *Visualized Centroid Values for each Attribute*

The plot tab selected in Figure 8-14c is the plot of the centroid table which helps to see which attributes describe different clusters.

1. *Interpret your results*: To characterize the resulting clusters, we examine the cluster centroids, using this line chart (also known as the profile plot). We can see from Figure 8-14c that the clusters are not too separated from each other for some attributes, but nonetheless there are some attributes that give distinct separation such as the Brand = Dangote, Brand = Chikki, Product class = medium, Product class = low, Product class = high, Product size = large, Product size = small, and list price. These are the main distinguishing attributes for this k-means model, and they can be used to characterize each of the clusters.

To characterize the different clusters, compare the cluster centroid (centroid table in Figure 8-14c) and try to give each cluster a label. This label is obtained from any or all of the variables in the dataset.

The question here then is: Which clusters would you target for offers, and what types of offers would you target to customers in that cluster?

8.4 Association Analysis

Association analysis is a technique used to discover groups of related items (clusters) in databases that store transactions.

It aims to figure out what kinds of things people tend to buy together. When this discovery is made, these items can be shown together among the other items, offered in post-transaction coupons, or recommended for online shopping. In order to select a subset of rules, the stages of establishing association rules include formulating rules and evaluating the strength of those rules.

Algorithms for mining frequent itemsets are used to construct association rules. The classic one selected is the Apriori algorithm of Agrawal et al. (1993). The important steps in the algorithm are shown in Figure 8-15.

First generate frequent itemsets with just one item (one-itemsets)
Then recursively generate frequent itemsets with two items
Then with three items, and so on until we have generated frequent itemsets of all sizes.
To generate frequent one-itemsets we count for each item, how many transactions in the database include the item.
These transaction counts are the supports for the one-itemsets.
We drop one-itemsets that have support below the desired minimum support to create a list of the frequent one-itemsets.
To generate frequent two-itemsets, we use the frequent one-itemsets. (The reasoning is that if a certain one-itemset did not exceed the minimum support, any larger size itemset that includes it will not exceed the minimum support.) etc.

Figure 8-15.   Apriori algorithm by Agrawal et al. (1993)

The algorithm can generate a lot of rules, but we only need to make a decision using the strongest based on the dependence between the antecedent and consequent. If we have the following rule:

If diaper and biscuits are purchased, THEN wipes are purchased on the same trip.

Antecedent: Diaper and biscuits are purchased.

Consequent: Wipes are purchased on the same trip.

The strength of association implied by a rule is measured using

$$Support = P(antecedent\ AND\ consequent) \qquad \text{(Eq. 8.1)}$$

The support for the rule indicates the rule's overall impact in terms of size: What is the total number of transactions impacted? If only a few transactions are impacted, the rule may not be of much benefit.

$$Confidence = \frac{no.transactions\ with\ both\ antecedent\ and\ consequent\ itemsets}{no.of\ transactions\ with\ antecedent\ itemset}$$

$$\text{(Eq. 8.2)}$$

$$Confidence = \frac{P(antecedent\ AND\ consequent)}{P(antecedent)} = P(consequent|antecedent)$$

$$\text{(Eq. 8.3)}$$

- A high value of confidence suggests a strong association rule (in which we are highly confident).

- However, this can be deceptive because when the antecedent and/or the consequent have strong support, we can have a high level of confidence in both even though they are independent.

- The confidence tells us the rate at which consequences will be discovered and is useful in judging a rule's usefulness in business or operational context. There may be too few consequences found by a rule with a low degree of confidence for it to justify the cost of promoting that rule's consequent in all transactions that include the antecedent.

$$lift\ ratio = \frac{confidence}{benchmark\ confidence} \qquad \text{(Eq. 8.4)}$$

- We may better evaluate an association rule's strength by comparing its confidence level to a benchmark value, in which we assume that the antecedent and consequent itemsets in a transaction are independent of each other. In other words, what level of confidence can we have in the results if the antecedent and consequent itemsets are independent?

- A lift ratio larger than 1.0 implies that the relationship between the antecedent and the consequent is more significant than would be expected if the two sets were independent. The larger the lift ratio, the more significant the association.

- The lift ratio is a measure of how well the rule is at finding the consequents, compared to random selection. Even though a very efficient rule is preferable to an inefficient rule, we must also take into account support: a rule with high efficiency but low support may not be as desirable as a rule with lower efficiency but much higher support.

Table 8-1 is the example of the format for the data used for association analysis using the Apriori algorithm. Each role represents the transactions, and each column represents an item, and the total column is the total items available. An example of the source of transactional data for association analysis is data collected using barcode scanners in supermarkets where each record lists all items bought by a customer on a single-purchase transaction.

Table 8-1. *An Example of Data Used for Association Mining*

Trans.	Crocs	KennethCole	LIbean	Nike	Adidas	Skechers	Fila
1	0	1	1	1	1	0	1
2	0	0	1	0	1	0	1
3	0	1	0	0	1	1	1
4	0	0	1	1	1	0	1
5	0	1	0	0	1	0	1
6	0	0	0	0	1	0	0
7	0	1	1	1	1	0	1
8	0	0	1	1	0	0	1
9	0	0	0	0	1	0	0
10	1	1	1	1	0	0	0
11	0	0	1	0	0	0	1

Association analysis – application for SMEs

The retail industry makes extensive use of association analysis to gain insight into the products that customers buy together. Knowing if particular sets of things are routinely purchased together is one type of choice that can be made. Association analysis aids in the design of catalogs and the layout of retail stores, as well as in cross-selling, promotions, and the identification of client categories based on purchasing habits. In order to increase SMEs' sales while reducing the costs of overstocking, the results of association analysis can be used to stock the shelves with the correct product mix. When it comes to cross-selling products, the results of association analysis can be used to increase the average spend per customer. SMEs' frontline employees are likely to have varying levels of expertise in this art because their experience and observational skills differ. Association analysis can help small businesses learn more about the kind of items their consumers are most likely to purchase so that they may better train and guide their employees in this area.

Association analysis example

A shoe retailer stores the data for the sales made to each customer and is able to convert it to the transactional data stored in shoes.xlsx. The intention of the shoe retailer is to analyze associations among purchases of these shoe items for purposes of point-of-sale display and guidance to sales personnel in promoting cross-sales.

1. Create a new repository (call it Association); *Import* the data named shoes.xlsx. When importing, make sure to convert all the columns to binominal data type apart from the Transaction no (Trans.). After importing the data, check the *Statistics* to be sure that there are no missing values.

2. Use the settings in Figure 8-16 to create frequent itemsets as displayed in the results in Figure 8-17 after running the process.

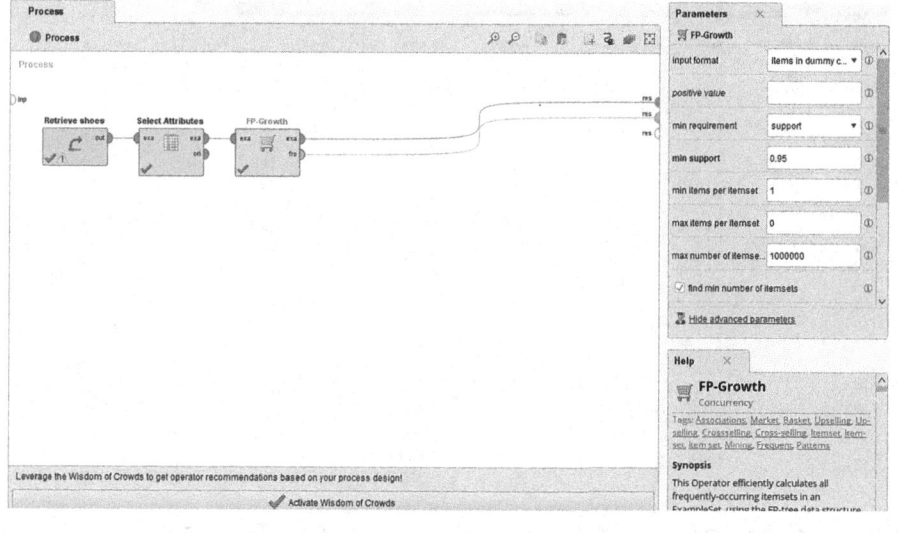

Figure 8-16. *Process to create frequent itemsets*

In Figure 8-16, for the ***Select Attributes*** operator, select all the binominal attributes apart from *Trans*. For the FP-Growth, leave the setting as default. Sometimes, depending on the data, you might need to reduce the support to get any result at all. The FP-Growth algorithm is an efficient algorithm for calculating frequently co-occurring items in a transaction database.

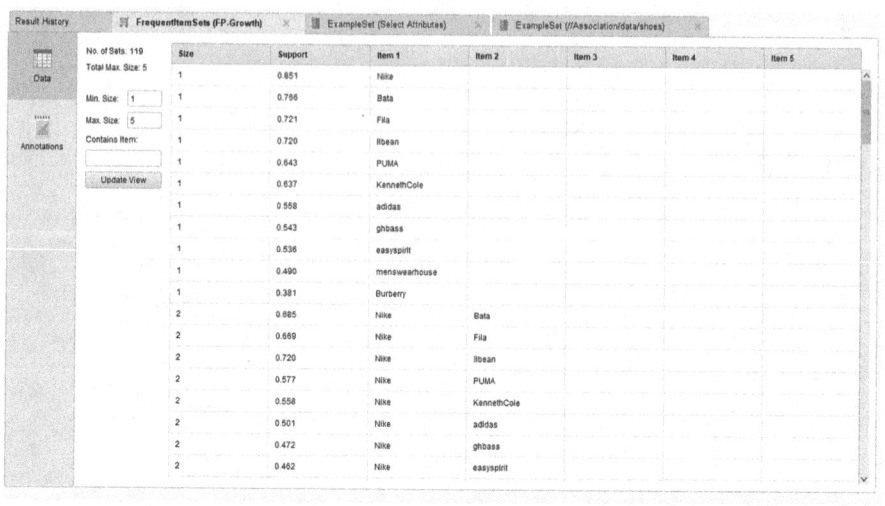

Figure 8-17. *Output of the process to create frequent itemsets*

Modify the settings in Figure 8-16 to give Figure 8-18 which is the complete settings for generating association rules. For now, in Figure 8-18, we leave the parameters for the ***Create Association Rules*** operator as default. But this can be adjusted based on the nature of the data.

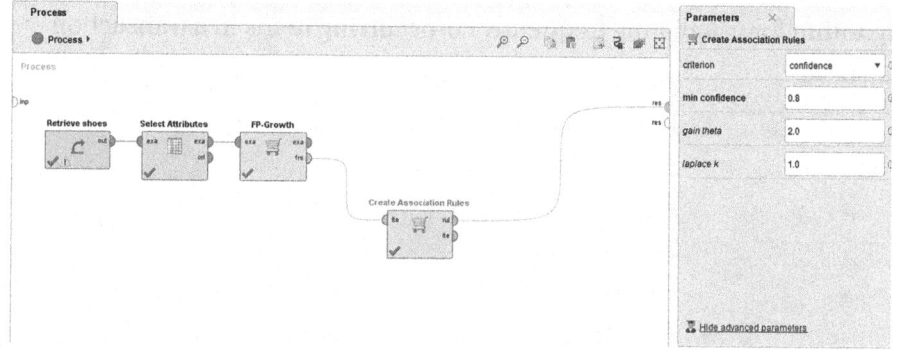

Figure 8-18. *Process to create association rules*

The result of running Figure 8-18 is given in Figure 8-19. It shows the premises (antecedent), conclusion (consequent), support, confidence, lift, and so on for the generated rules. To select the best to act upon, we are looking for the highest support, uplift at least greater than one with high confidence.

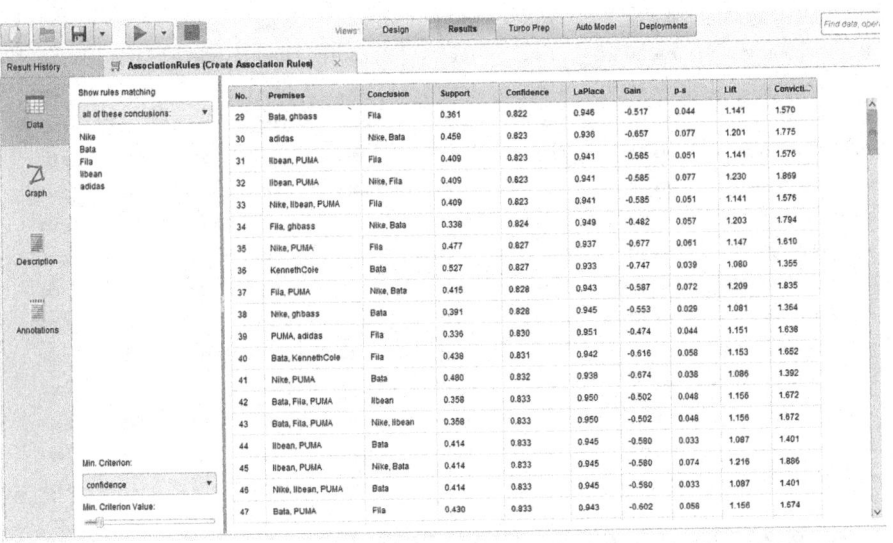

Figure 8-19. *Resulting association rule*

Sorting by lift, in order to select the best rules, we get Figure 8-20.

No.	Premises	Conclusion	Support	Confidence	LaPlace	Gain	p-s	Lift ↓	Convicti...
171	Fila, Ilbean, adidas	Nike, Bata	0.341	0.945	0.985	-0.381	0.094	1.379	5.686
150	Ilbean, adidas	Nike, Bata	0.392	0.918	0.975	-0.462	0.100	1.340	3.843
142	Fila, Ilbean, KennethC...	Nike, Bata	0.354	0.905	0.973	-0.428	0.086	1.322	3.329
112	Bata, Ilbean, adidas	Nike, Fila	0.341	0.870	0.963	-0.443	0.079	1.300	2.544
106	Bata, Ilbean, Kenneth...	Nike, Fila	0.354	0.866	0.961	-0.464	0.080	1.294	2.461
102	Bata, Ilbean, PUMA	Nike, Fila	0.358	0.865	0.960	-0.470	0.081	1.293	2.447
122	Fila, adidas	Nike, Bata	0.399	0.879	0.962	-0.509	0.088	1.283	2.600
119	Fila, Ilbean, PUMA	Nike, Bata	0.358	0.875	0.964	-0.460	0.078	1.278	2.526
116	Ilbean, easyspirit	Nike, Bata	0.342	0.870	0.963	-0.444	0.073	1.270	2.427
63	Ilbean, adidas	Nike, Fila	0.361	0.845	0.954	-0.493	0.075	1.264	2.141
58	Bata, Ilbean	Nike, Fila	0.489	0.845	0.943	-0.669	0.102	1.262	2.129
97	Fila, Ilbean	Nike, Bata	0.489	0.864	0.951	-0.643	0.101	1.261	2.315
74	Ilbean, KennethCole	Nike, Bata	0.409	0.852	0.952	-0.551	0.080	1.244	2.130
67	PUMA, adidas	Nike, Bata	0.343	0.847	0.956	-0.467	0.066	1.236	2.058
170	Fila, Ilbean, adidas	Bata	0.341	0.945	0.985	-0.381	0.064	1.233	4.224
172	Nike, Fila, Ilbean, adid...	Bata	0.341	0.945	0.985	-0.381	0.064	1.233	4.224
56	Fila, easyspirit	Nike, Bata	0.340	0.844	0.955	-0.466	0.064	1.232	2.015
32	Ilbean, PUMA	Nike, Fila	0.409	0.823	0.941	-0.585	0.077	1.230	1.869
166	Nike, Fila, adidas	Bata	0.399	0.939	0.982	-0.451	0.073	1.226	3.825

Figure 8-20. *Sorting Figure 8-19 by lift ratio*

- Interpreting the first rule in Figure 8-20, we have

If Fila, Ilbean, and Adidas are purchased together, then, with confidence of 94.5%, Nike and Bata will also be purchased. This rule has a lift ratio of 1.379.

The preceding rule can be used in several ways, for example, if the customer is requesting for a particular brand of shoe, you can introduce a new brand to the customer. This new brand is recommended based on solving the brand association problem which makes it certain that the customer will also want to buy the new brand. This type of inference can be implemented in several ways by SMEs.

Examples include through email marketing, recommender engines, promotions, and so on.

8.5 Network Analysis

A network is a set of objects (nodes) with interconnections (edges). An example of a communication network obtained from tweets is shown in Figure 8-21.

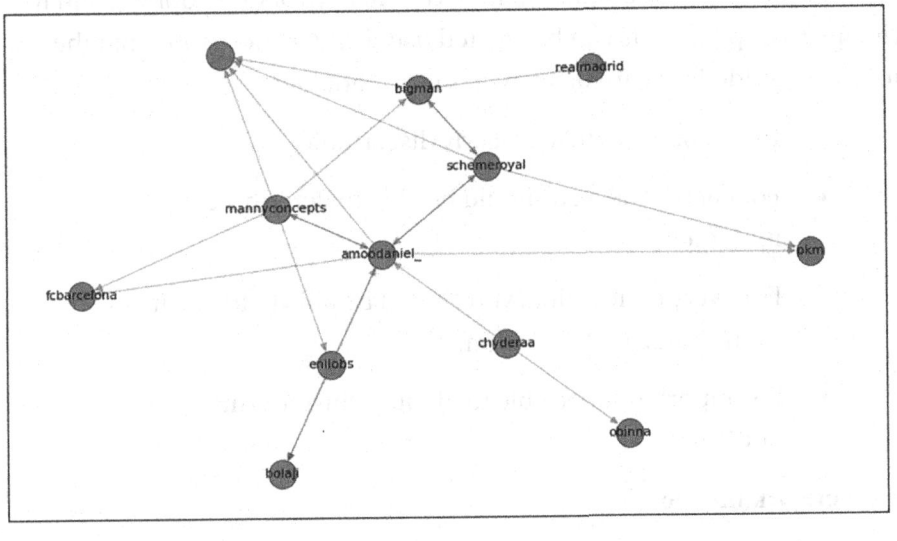

Figure 8-21. *An example of a social communication network*

There are different types of networks. There are social networks that are created from social interactions and personal relationships, for example, on social media, co-authorship networks, and so on. There are other types of networks such as transportation and mobility networks, information networks, biological networks, and so on.

The nodes in the network refer to objects such as people, departments, transport stations, and so on. The edges refer to the connection between the objects which could be friendship, marriage, family ties, research communication, and so on. Weighted edges can also reflect connection qualities, such as how much email is exchanged between two people on a network. Distance between two points can be depicted using the length of an edge on a map.

Networks can either be undirected (bidirected) or directed in structure. An example of a directed network is if Ibk follows Femi on Twitter, and Femi does not follow Ibk. An example of an undirected or bidirected network is if two people co-author a research publication together.

There are various tools and platforms from which we can create and visualize networks; examples include Gephi, networkX, Graphviz, and R (igraph). Graph layouts can be created using any of these tools, but the following guidelines can make them more readable:

- Each node should be clearly discernible.

- For each node, you should be able to determine its degree.

- For every connection, you should be able to follow it from source to destination.

- It's important to be able to identify clusters and outliers.

Network metrics

Degree distribution: In a network, the degree distribution describes how many nodes a node is connected to. It helps to know the number of nodes in the network and the range of their edges.

Density: This gives the overall connectedness of a graph that focuses on the edges, not the nodes. It is the ratio of the actual number of edges to the maximum number of potential edges.

Average clustering coefficient: This is also known as "triadic closure." The clustering coefficient indicates the possibility that a node that shares connections in the network would join the network.

Network diameter: This is used to describe the maximum distance between any pair of nodes in the network.

Number of communities: This gives the number of the subset of nodes in the network, such that every author in the subset has a path to every other author and no other author has a path to any author in the subset.

Degree centrality: This tries to find centrality based on the idea that significant nodes have a large number of connections.

Closeness centrality: This aims to find centrality by assuming that significant nodes will be close to all other nodes in a network.

Betweenness centrality: The search for centrality is founded on the notion that nodes that connect to other nodes are significant nodes.

Eigenvectors: To convey the idea that a node is more central if its neighbors are more central.

For more on network metrics, check Tsvetovat and Kouznetsov (2011).

Network data

Network data can either be in the form of the edge list or the adjacency list. Figures 8-22a and 8-22b present the edge list and adjacency list (of the same data), respectively.

A	B
A	C
A	D
A	E
A	F
B	A
B	C
B	E
B	H
C	A
D	G
E	A
F	H
G	F
H	D

Figure 8-22a. *Edge list*

	A	B	C	D	E	F	G	H
A	0	1	1	1	1	1	0	0
B	1	1	0	0	1	0	0	1
C	1	0	0	0	0	0	0	0
D	0	0	0	0	0	0	1	0
E	1	0	0	0	0	0	0	0
F	0	0	0	0	0	0	0	1
G	0	0	0	0	0	1	0	0
H	0	0	0	1	0	0	0	0

Figure 8-22b. *Adjacency list*

Network analysis can be applied in almost all of the domain of interest. In business, we can use it to answer questions like the following:

- Is an advertisement likely to spread in the network?

- Who are the most influential people in this business environment?

- What products are highly connected to other products?

- And many more.

In this section, we will use two simple examples to demonstrate how to create networks and customize them in Gephi, and in Section 8.6, we will demonstrate a real business problem scenario example of the application of network analysis. Note that this can be customized to solve several related problems also. For these examples, Gephi will be used for the practical demonstration. Section 1.8 of this book gives the details of how to set up Gephi and to be familiar with the Gephi environment.

Network analysis – example 1

The data named *dynamic.gexf* is a Gephi file that contains tweets from users in a communication social network (Twitter). These tweets have been converted to an edge list in the form of a Gephi file (data was downloaded and converted using netlytics.org). For a complete tutorial of how to download the data and create the Gephi file which contains the nodes and edges, please visit https://yusufsalman.medium.com/twitter-network-analysis-and-visualisation-with-netlytic-and-gephi-0-9-1-1011b009261. The downloaded data in Excel is presented in Figure 8-23.

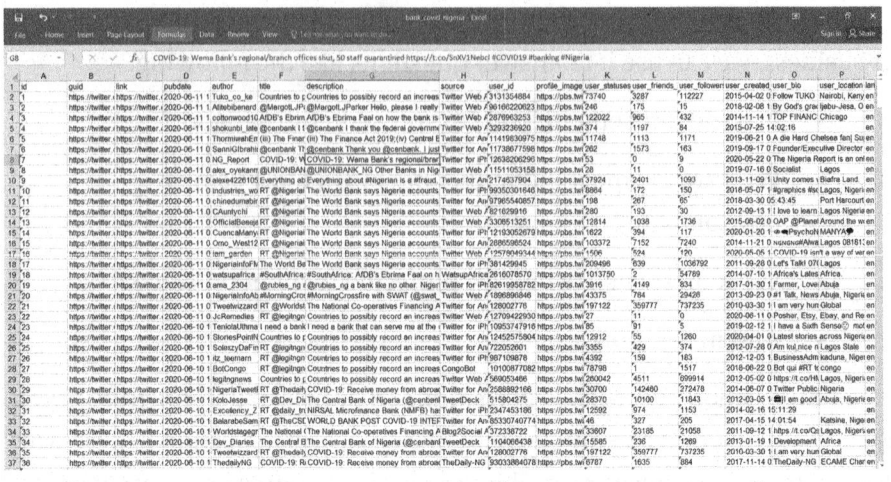

Figure 8-23.   *Downloaded data in Excel format*

1. Load the Gephi application as seen in Figure 8-24.
   Click ***Open Gephi file*** and select **dynamic.gexf** to
   reveal the interface in Figure 8-25.

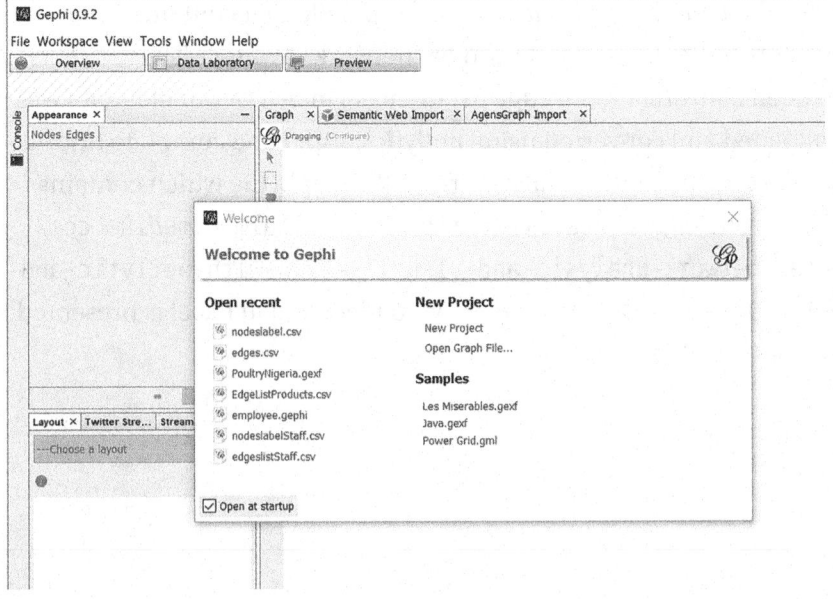

Figure 8-24.   *Loading the dynamic.gexf file*

In Figure 8-25, you can set the **Graph Type**, which in this case is directed (by default). The number of nodes and edges is also revealed on this interface, together with some other details. Figure 8-25 also asks if you want to open the file in a new workspace or append it to an existing workspace. If you are importing the node list which is a detailed description of the nodes, the option of append will be used, but for this example, we will open in a new workspace. Click **OK**.

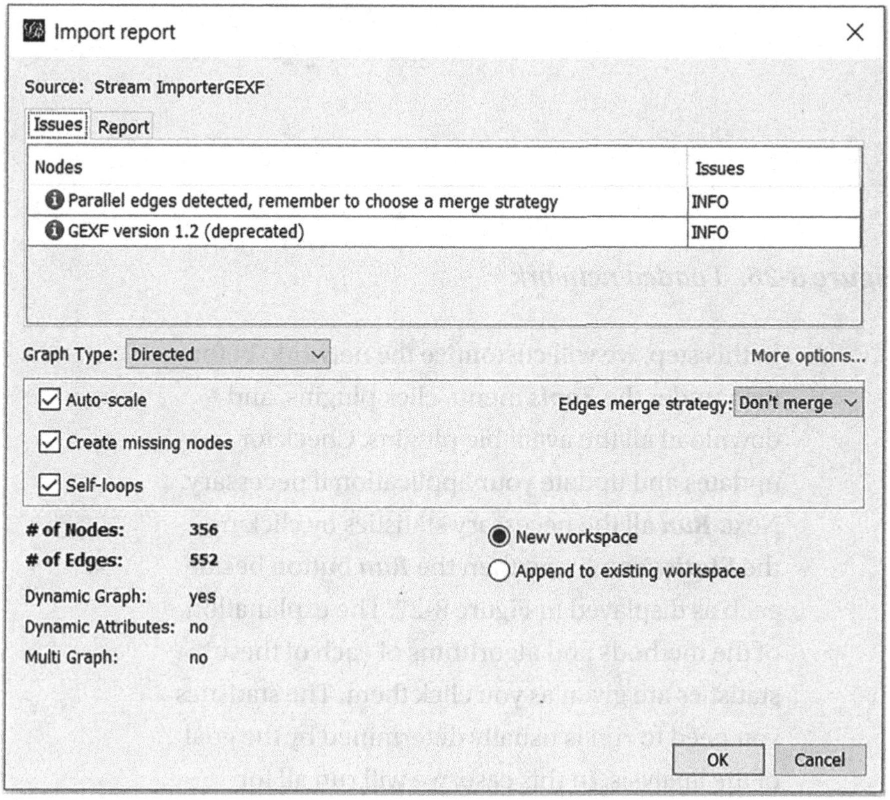

Figure 8-25. *Importing the interface*

The network created is then displayed in
Figure 8-26.

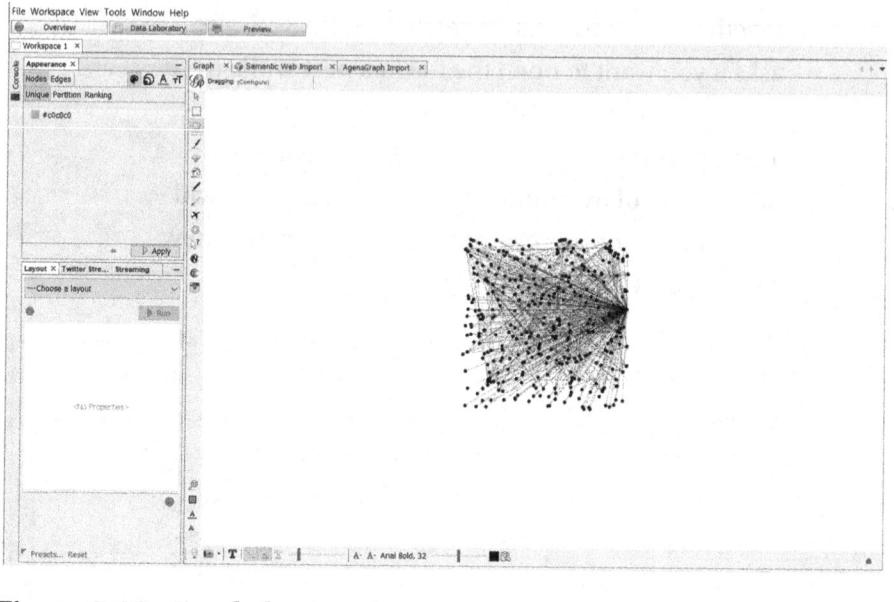

Figure 8-26. *Loaded network*

2. In this step, we will customize the network. Before
    that, under the ***Tools*** menu, click plugins, and
    download all the available plugins. Check for
    updates and update your application if necessary.
    Next, ***Run*** all the necessary statistics by clicking
    the ***Statistics*** tab and then the ***Run*** button beside
    each as displayed in Figure 8-27. The explanation
    of the methods and algorithms of each of these
    statistics are given as you click them. The statistics
    you need to run is usually determined by the goal
    of the analysis. In this case, we will run all for
    demonstration purposes.

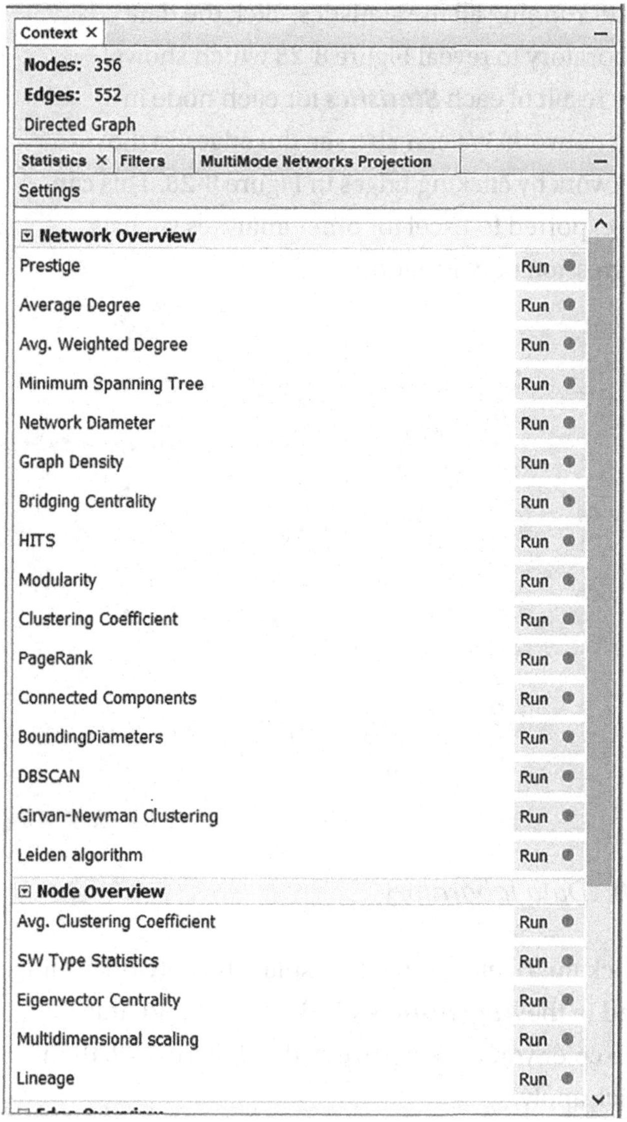

Figure 8-27. *Statistics in Gephi*

3. After running all the statistics, click the data
    laboratory to reveal Figure 8-28 which shows
    the result of each *Statistics* for each node in
    the network. We can also see the edges in the
    network by clicking Edges in Figure 8-28. This can
    be exported to Excel for other analyses such as
    regression or correlation.

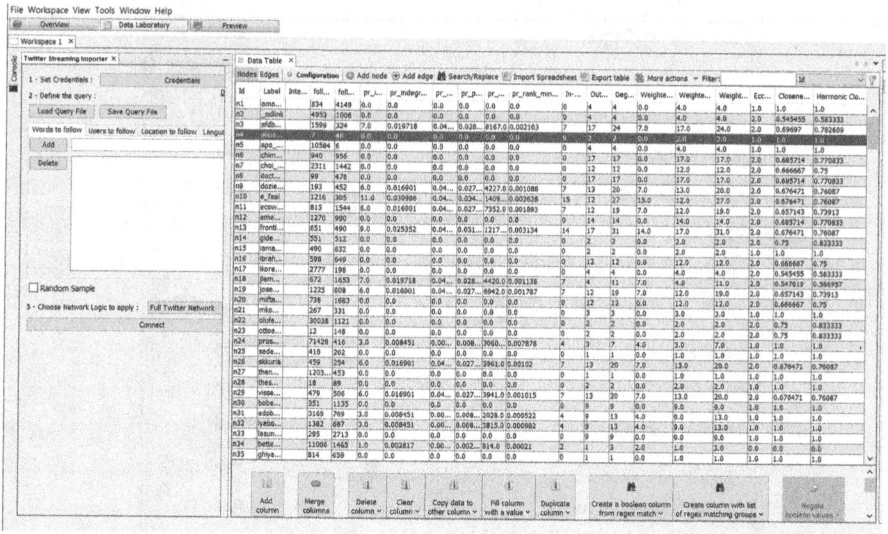

Figure 8-28. *Data laboratory*

4. Click the *Overview* tab to display the network again,
    and in the *Appearance* window, click and drag
    the color you want as revealed in Figure 8-29, then
    click apply.

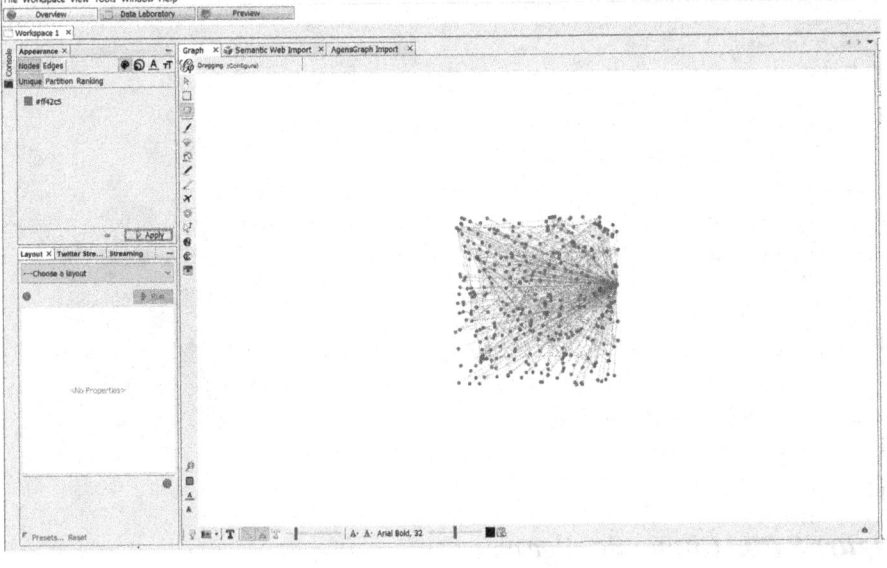

Figure 8-29. *Changing node color*

5. Use the ***Edges*** tab of the Appearance window to add
    color to the edges (Figure 8-30) and adjust the sizes
    of nodes using eigenvector centrality (Figure 8-31).
    This can be done by selecting the nodes in the
    Appearance window and selecting size and ranking
    as seen in Figure 8-31. The importance or influence
    of units (nodes) in a graph is measured by their
    centrality. We choose the eigenvector centrality in
    this demonstration.

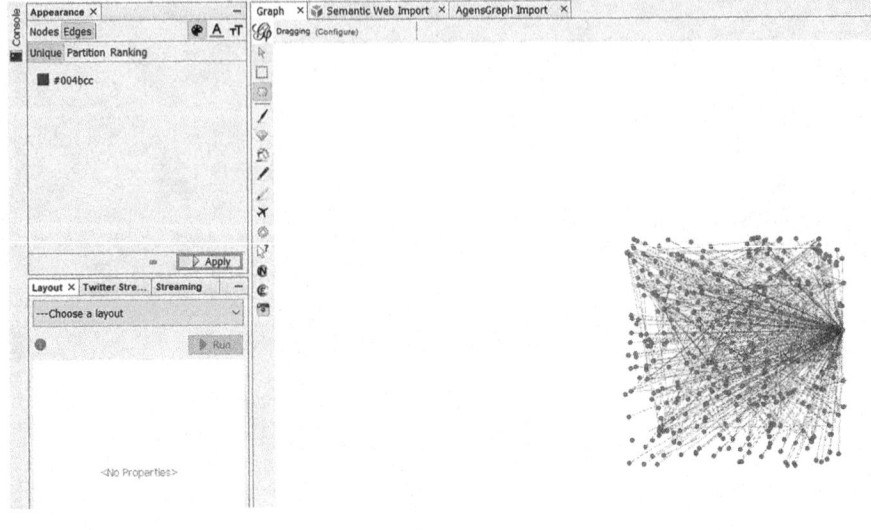

Figure 8-30. *Changing edge color*

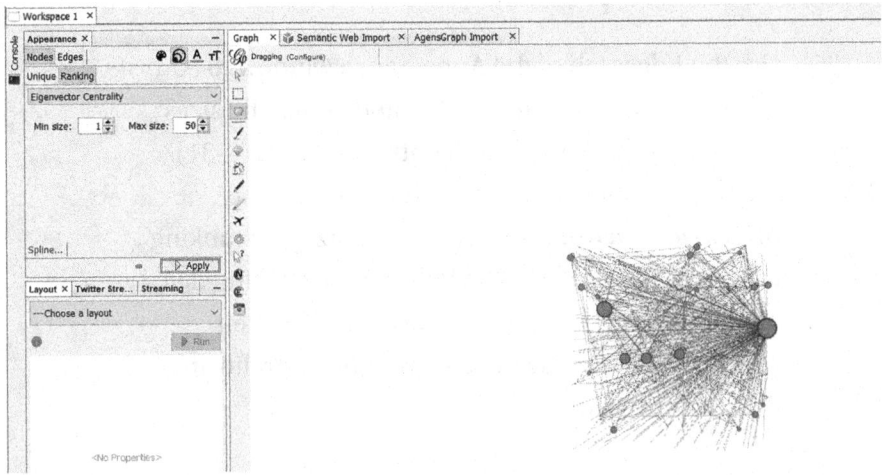

Figure 8-31. *Ranking the edges using eigenvector centrality*

6. Gephi offers various different layouts to present your
    graph and a few plugins to make it look better. For
    this example, we will use one of the most popular
    layouts: ***Fruchterman-Reingold***. It's a force-
    directed graph algorithm, and it will give us a clear
    view of the connections in our graph. Figure 8-32
    shows the setting of running this layout.

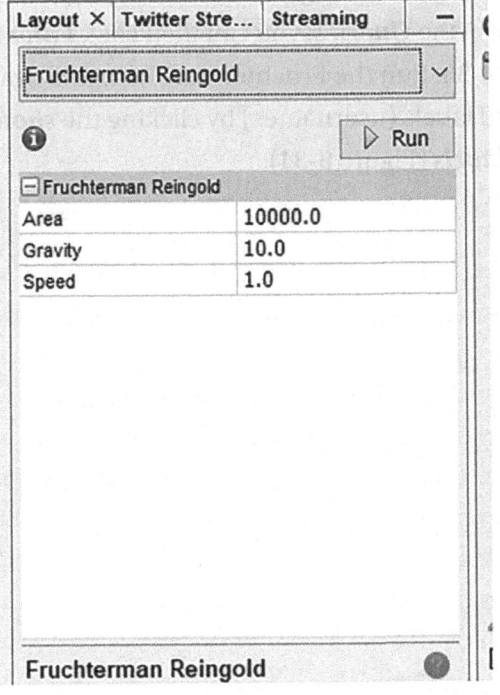

Figure 8-32. *Fruchterman-Reingold layout*

7. Click **Run** in Figure 8-32, stop after a few seconds, and eliminate all nodes with less than three connections (use the settings in Figure 8-33). This will tidy up the graph and allow us to see more important edges (while not affecting the statistics). On the right side of the window, click the **Filters** tab, and click **Attributes** under **Filter** and **Range** under **Attribute**. Under **Range**, double-click **Degree**. It will appear in the **Queries**; you can then click **Filter** in Figure 8-33. Run the Fruchterman-Reingold again and add labels (usernames) by clicking the **show Node labels** (Figure 8-34).

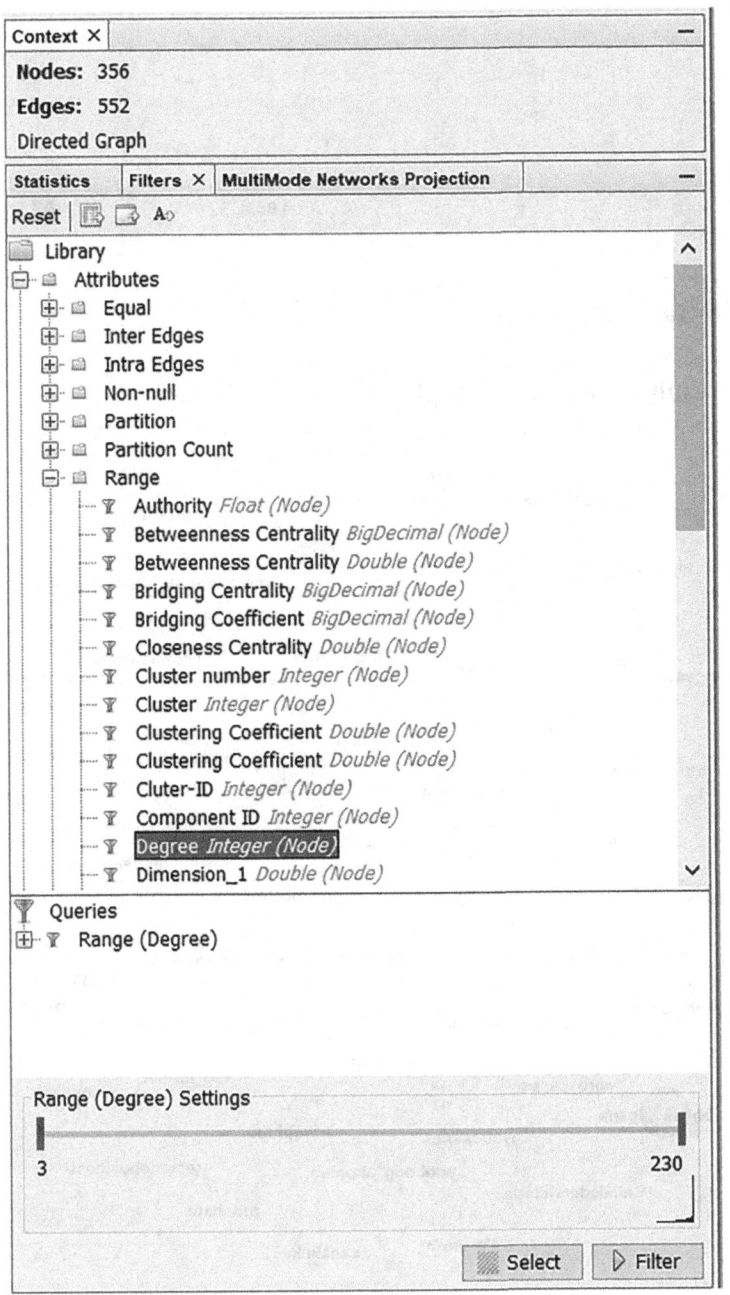

Figure 8-33. *Filtering the degrees*

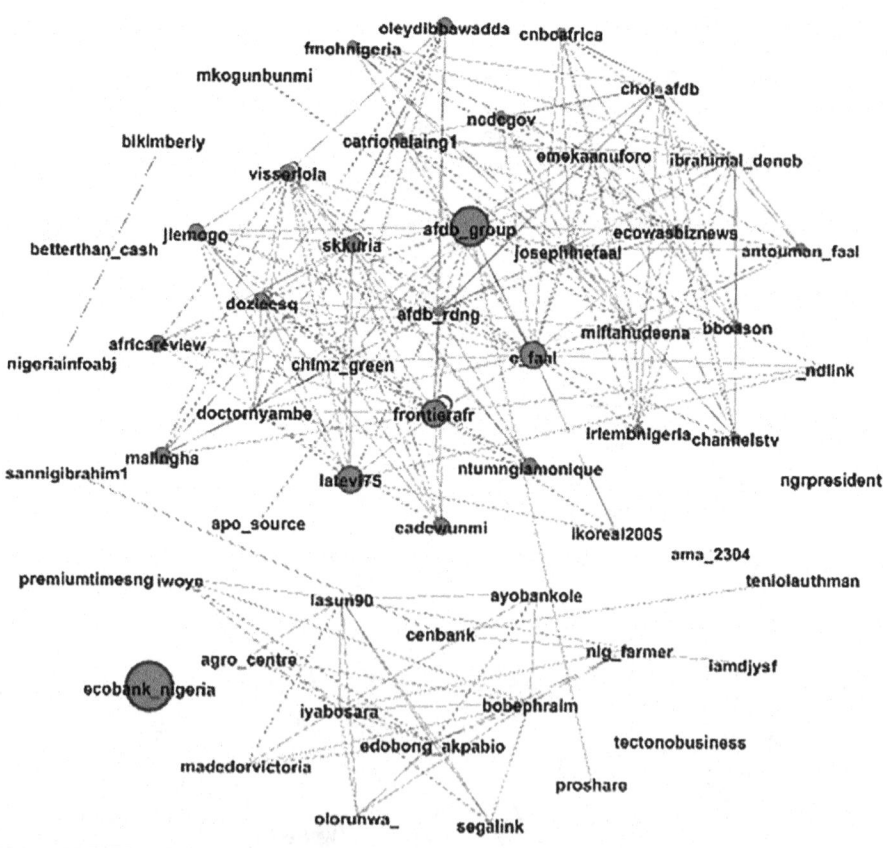

Figure 8-34. *Show labels*

The resulting network is displayed in Figure 8-35.

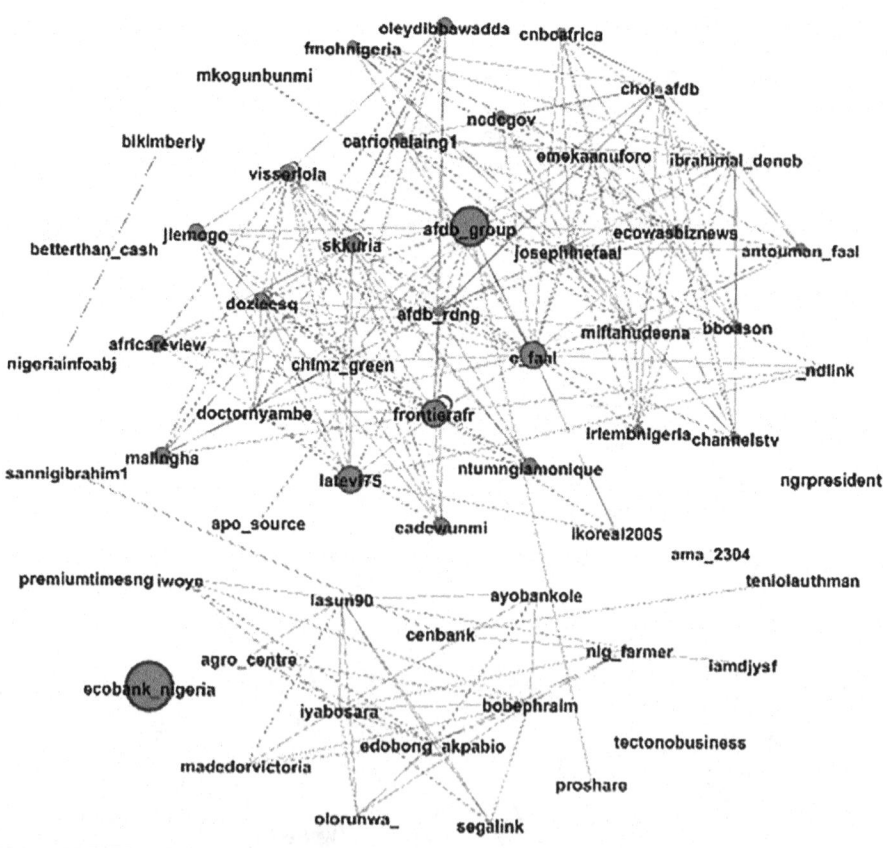

Figure 8-35. *Network of Users*

For this example, we will stop here, but there are a lot of other settings that can be used to visualize this network for various types of insights. Also, the preview tab can be used for better visualization of the network. From Figure 8-35, we can see the node ecobank_nigeria has the highest eigenvector centrality, signifying that it is the most central because its neighbors are also central.

Network analysis – example 2

For the second example, we will be creating a product network of a food and agricultural business. The edge list is created from the associations of the products as they are purchased together. The data file to be used for creating this network is named EdgeListProducts.csv and NodeListProducts.csv, respectively.

1. Open a new project in Gephi. Load the data into Gephi as an undirected graph (File ➤ Open), and load the EdgeListProducts.csv first. Figure 8-36 is the interface after successfully loading the edge list into Gephi.

Figure 8-36. *Successfully loading EdgeListProducts.csv*

2. Load the node list by clicking ***File*** ➤ ***Import spreadsheet***, select the NodeListProducts.csv, click Next, and then click Finish. On the resulting interface in Figure 8-37, make sure to change the ***Graph Type*** to ***undirected*** and select the ***Append to existing workspace***.

3. Run all the ***Statistics***. Filter out nodes with a degree not up to 12.

4. Change the node to pink and the edge to blue, and use ***eigenvector centrality*** to resize the nodes (just as in example 1); we should have Figure 8-37.

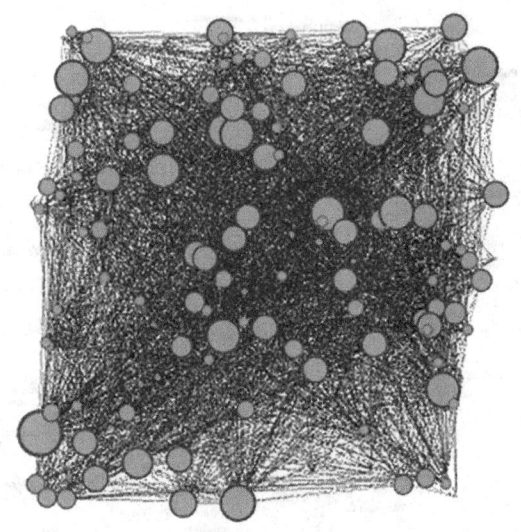

Figure 8-37. *Resizing the graph using eigenvector centrality*

5.   Run the ***Fruchterman-Reingold*** layout to see
     Figure 8-38, click the ***Show label*** button, run the
     Label Adjust layout, and run the ***Expansion*** layout
     to resize the network so that it is bigger and labels
     are readable. The result should look like Figure 8-38.

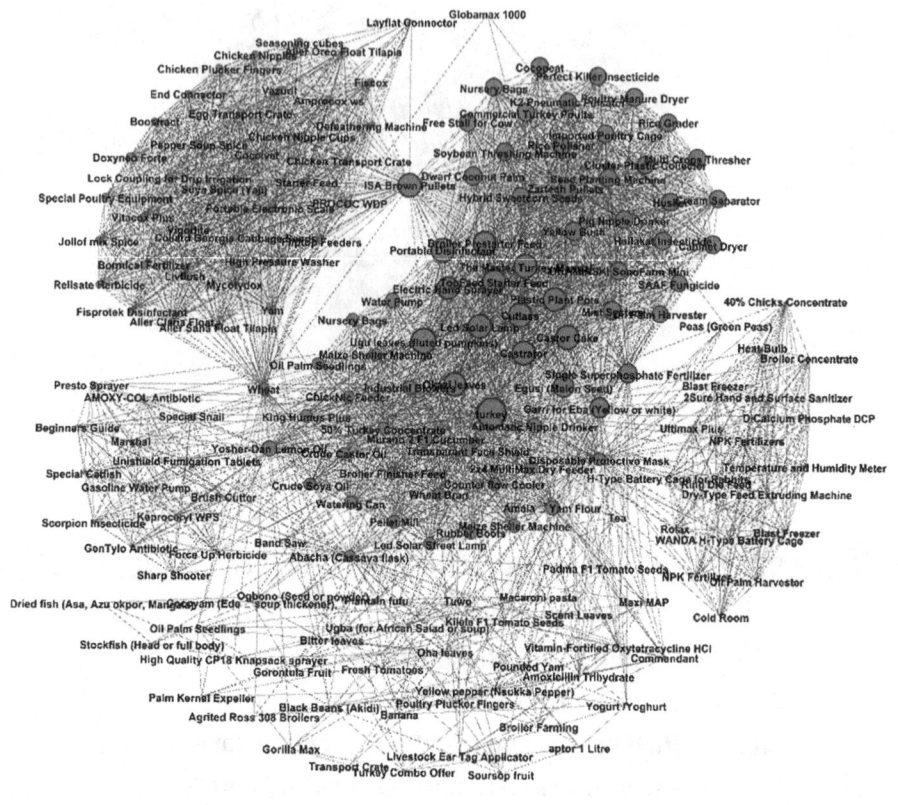

Figure 8-38. *Network of products (eigenvector centrality)*

The network graph in Figure 8-38 is the result of using the eigenvector centrality to rank the nodes on the network. From Figure 8-38, we are able to see nodes which have a high centrality value because they are connected to nodes that are also central. A high eigenvector means that a node is connected to nodes that have high eigenvector scores.

1. Using ***betweenness centrality***, we get the network in Figure 8-39.

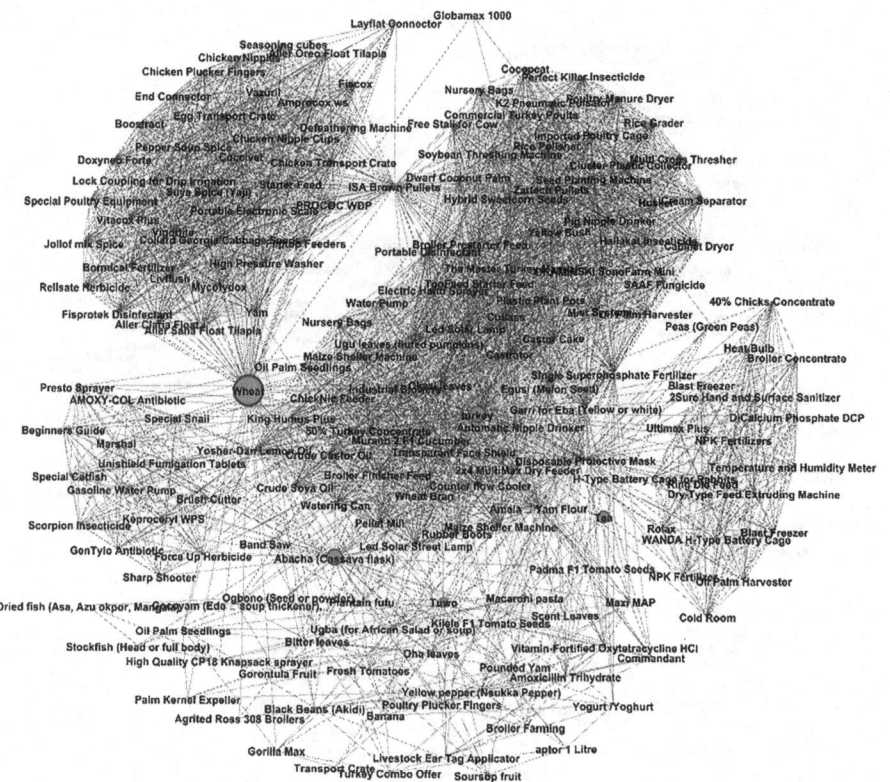

Figure 8-39. *Network of products (betweenness centrality)*

The network graph in Figure 8-39 is the result of using the betweenness centrality to rank the nodes on the network. From Figure 8-39, we are able to see that nodes with a high betweenness centrality form a bridge between other clusters of nodes.

2. Using the ***closeness centrality***, we get the network in Figure 8-40.

Figure 8-40. *Network of product closeness centrality*

The network graph in Figure 8-40 is the result of using the closeness centrality to rank the nodes on the network. From Figure 8-38, we are able to see nodes which have a high centrality value because they are close to other nodes in the network. References for the application of these centrality metrics in different fields can be found in Section 8.8 of this chapter.

8.6 Practical Business Problem V (Staff Efficiency)

Collaboration networks have been used in people analytics to reduce employee overload, improve the resiliency of global teams, enhance career paths, and so on. The following are some of the metrics used in collaboration networks:

Network size: Number of nodes attached to a node. In-degree (measure the number of nodes connected from other nodes) and out-degree (measure the number of nodes the particular node is connected to). This is captured by an average degree in Gephi.

Network strength: Strong ties and weak ties. How weak or strong the in-degree or out-degree is. This can be captured by a weighted degree in Gephi.

Network range: How many different types of people you are connected to. A high range network has the ability to get information from a variety of sources. This can be investigated by checking how many communities you belong to.

Network density: Measures the degree of interaction between all members of a population and all other members. It is the overall interaction measure that measures how closely networked network members are. It is not calculated for each member. This is captured by Graph density in Gephi.

Other metrics are as explained in Section 8.5.

In this problem scenario, we will use network analytics for people analytics, one of the major application types of Business Analytics. We will particularly be focusing on analyzing collaboration networks. In the problem scenario, we have an organization that is looking at improving the efficiency of its employees. Even though this could be done in several ways, we will explore the possibility of using a communication network of the employees of the organization and investigate performance using this network.

In this problem scenario, the network data was collected via a survey and converted to a node (nodeslabelStaff.csv) and edge list (edgeslistStaff.csv), respectively.

In this problem scenario, we are looking at the possibility of generating any or some of the following insights. Is the network good or bad, are people really doing what they are supposed to be doing in terms of collaboration, and so on? To do this, we will answer questions like: How do collaboration patterns vary? How do collaboration patterns matter for important outcomes? This can be explored with respect to Individuals, groups, or organizations.

To answer the question, "How do collaboration patterns vary?", we can use simple descriptive statistics; we can pick a network size, for example, and compare across individuals and compare changes over time. The result of this can be used for managing employees in the following areas: performance assessment, roles and responsibilities, pay and promotions, training and mentoring, job rotations and career development, and retention.

To answer the question, "How do collaboration patterns matter for important outcomes we care about for our employees?", first we should realize that there are different types of outcomes for individuals such as performance (sales per quarter, cost savings, self-reported 1–3 ratings, manager-reported 1–3 ratings, bonus), satisfaction, commitment, burnout, turnover, etc.

In this practical example, we have chosen to use network size to examine performance. To do this, we will check the correlation between size and performance. For our case study, the performance data is captured in sales per quarter (measured in millions).

Practical demonstration

1. Load the edgeslistStaff.csv into Gephi, and set the **Graph Type** to **undirected**. Import the second csv file (nodeslabelStaff.csv) using **Import spreadsheet** under the **File** menu. When importing

nodeslabelStaff.csv, make sure to **append to
existing workspace**. Figure 8-41 shows the resulting
interface after this step is successfully done.

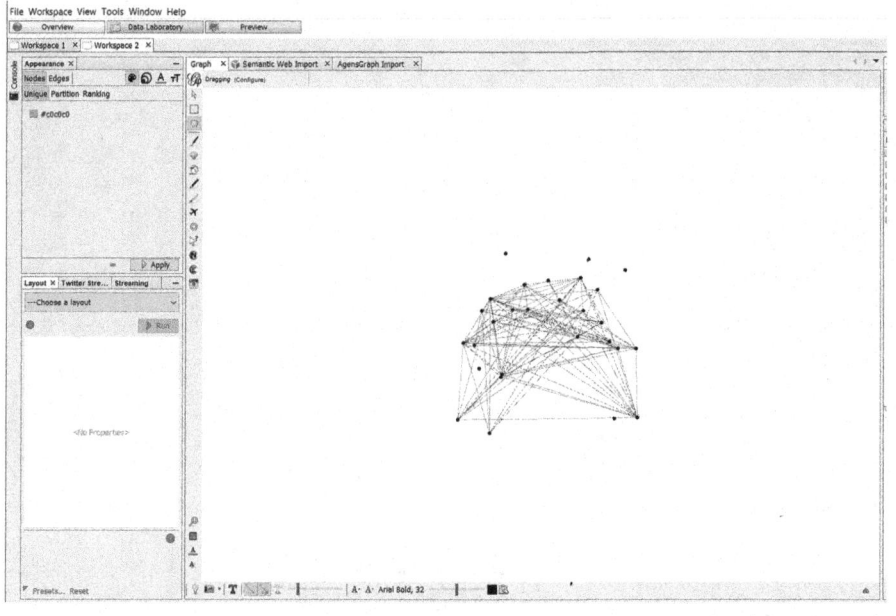

Figure 8-41. *Importing edgeslistStaff.csv*

2. Run Fruchterman-Reingold and then the expansion
    to give Figure 8-42.

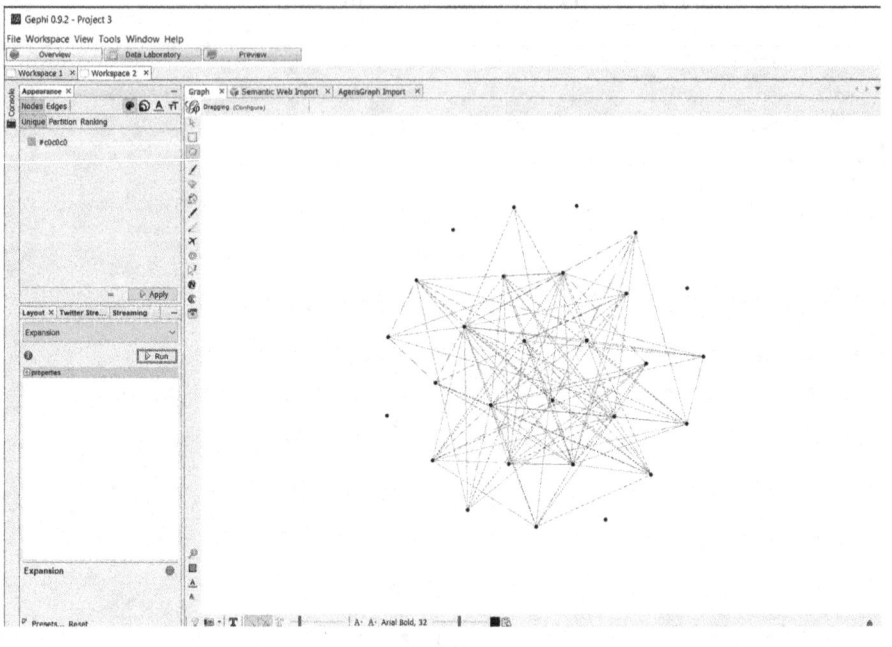

Figure 8-42.   *Expanding the network*

3. Run the ***Statistics*** (Prestige, Average Degree, Avg.
    Weighted Degree, Network Diameter, Graph density,
    Modularity, Connected Components, Eigenvector
    centrality). In the ***Appearance*** window, click **size**
    ➤ **Ranking** and choose attribute degree-min-1,
    max-50. (You can try several variations and see the
    one that suits your purpose).

4. Click ***show node labels***; adjust the size of nodes
    to make it bigger. There are so many things you
    can explore in the ***appearance***, ***layout***, and
    statistics windows to have a beautiful and attractive

network and also communicate different insight. As an example, we can visualize the clusters in the network by using the settings **Nodes ➤ Color ➤ Partition ➤ Modularity**. Click **Preview** (show node label in the preview); the final network should look like Figure 8-43.

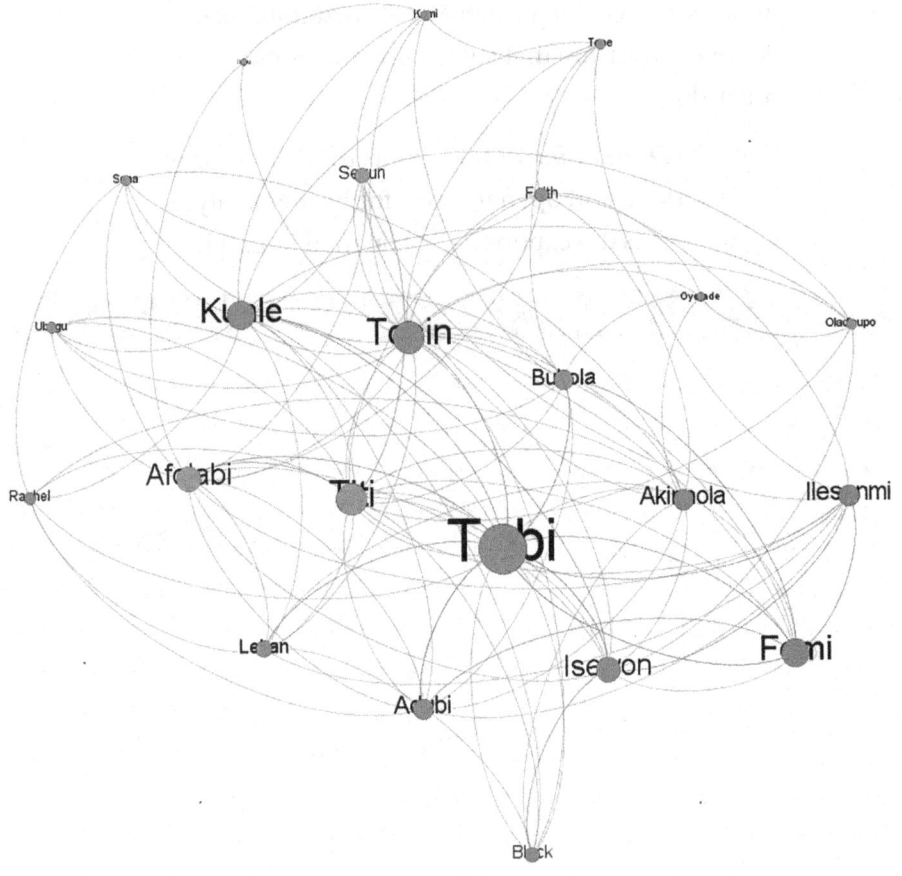

Figure 8-43. *The final staff communication network*

The color of the node indicates the cluster that each node belongs to, while the sizes indicate their degrees. The network graph in Figure 8-43 shows that the node *Tobi* has the highest degree in the organization's communication network; this indicates that a lot of nodes in this organization communicate with Tobi. The implication of this, for example, is that in such an organization, there is a need to ensure that such a node does not break down.

5. Click the **Data Laboratory**, export the **Nodes table** in csv format, and open in Excel for further analysis. Figure 8-44 is the exported data opened in Excel.

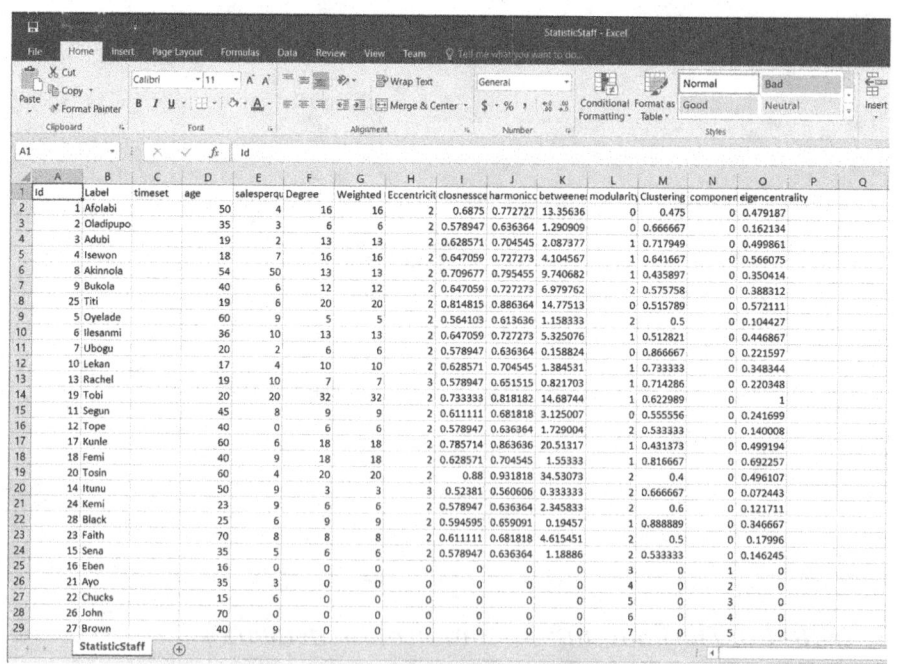

Figure 8-44. *Exported data in Excel*

6. Use correlation analysis in Excel to correlate the degree (network size) with performance. To do this, hold the **ALT key** and **type t** and **type I** in succession, and you will see the window to add the data analysis tool in Figure 8-45.

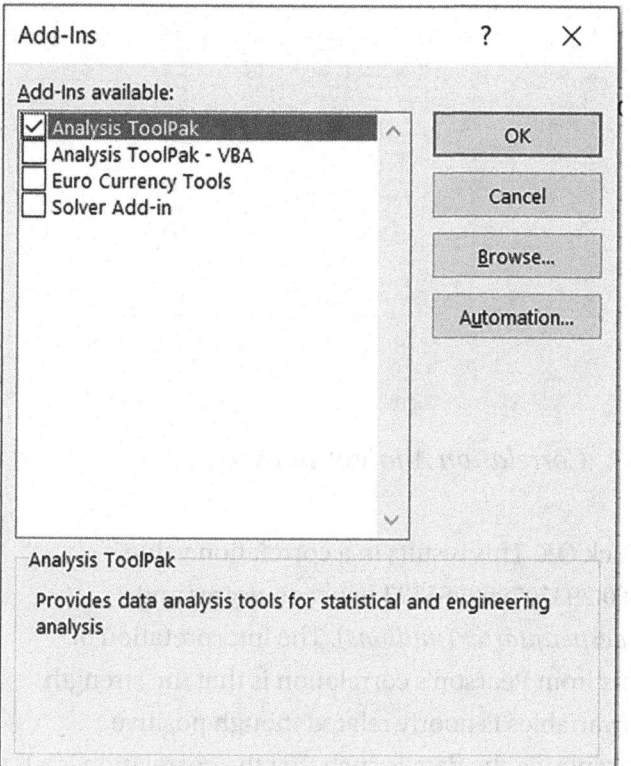

Figure 8-45. *Data analysis in Excel*

7. Click **OK**; click **Data Analysis** on the tools bar, select
    correlation, and select the input range and new
    workbook as the output option (Figure 8-46).

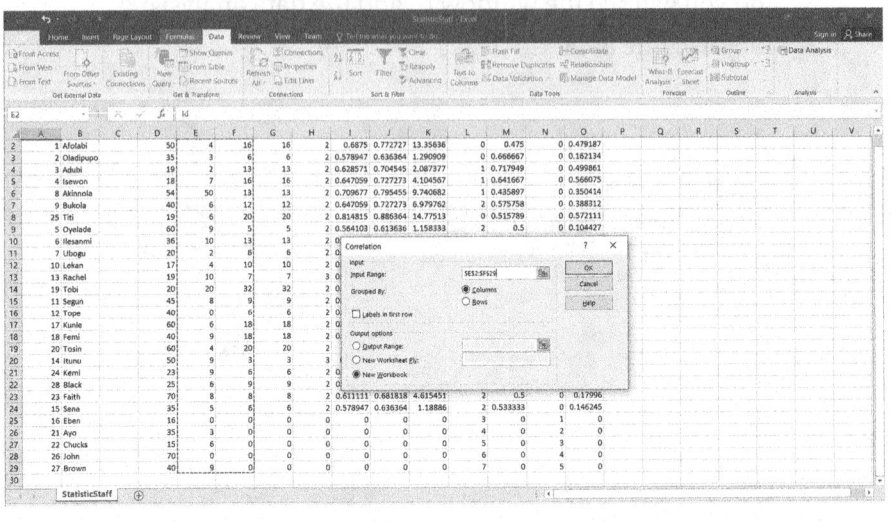

Figure 8-46. *Correlation Analysis in Excel*

8. Click **OK**. This results in a correlation value
    of 0.293707089965323 between *degree* and
    salesperquarter (millions). The interpretation of
    this from Pearson's correlation is that the strength
    of variables is poorly related though positive.
    (Assuming the data is such that the correlation is +1
    to 0.7, then the strength of the variables is perfectly
    related in a positive way, and we can then go ahead
    to look at how to increase the collaborations by
    providing motivations to build ties.)

8.7 Problems

1. Repeat the practical business problem V, using
    the same data (nodeslabelStaff.csv); this time
    around, correlate the centrality measures with the
    performance.

2. Using the dataset VegetableTransactions.csv,
    perform an association rule analysis, and comment
    on the results. Select the most important rules
    using the necessary criteria and give appropriate
    interpretation to the selected rules.

3. Using the data named Alegra.xlsx, perform
    clustering analysis using the k-means clustering
    algorithm. Validate, visualize, and interpret the
    resulting clusters.

8.8 References

1. Agrawal, R., Imielinski, T. and Swami, A. (1993)
    Mining Association Rules between Sets of Items
    in Large Databases. Proceedings of the 1993 ACM
    SIGMOD International Conference on Management
    of Data, Washington DC, 26-28 May 1993, 207-216.

2. Maksim Tsvetovat, Alexander Kouznetsov (2011)
    Finding Connections on the Social Web, Social
    Network Analysis for Startups.

3. Anil K. Maheshwari (2015) Business Intelligence
    and Data Mining, published by Business Expert
    Press, LLC.

4. Jiawei Han, Micheline Kamber, and Jian Pei (2012) Data Mining Concepts and Techniques, Third Edition, published by Elsevier.

5. Galit Shmueli, Nitin R. Patel, & Peter C. Bruce, Data Mining for Business Analytics, Concepts, Techniques and Applications in R, published by John Wiley & Sons, Inc., Hoboken, New Jersey, 2018.

6. Divyanshu Anand (June 2020) Gower's Distance, `https://medium.com/analytics-vidhya/gowers-distance-899f9c4bd553`

7. Rahul Kaliyath (July 2020) Day 12: Cluster Validation and Analysis, `https://medium.com/@rahulkaliyath/day-12-cluster-validation-and-analysis-f66f9b2bc384`

8. W. M. Rand (1971). "Objective criteria for the evaluation of clustering methods." Journal of the American Statistical Association. American Statistical Association. 66 (336): 846–850. doi:10.2307/2284239.

9. Chapter 17 Clustering Validation. `www.dcs.bbk.ac.uk/~ale/dsta/2020-21/dsta-8/zaki-meira-ch17-excerpt.pdf`

More resources on the chapter for further reading

- Yusuf Salman (April 2016) Twitter Network Analysis and Visualisation with Netlytic and Gephi 0.9.1, `https://yusufsalman.medium.com/twitter-network-analysis-and-visualisation-with-netlytic-and-gephi-0-9-1-1011b009261`

- Kmeans Practical Problem I, www.youtube.com/watch?v=H5O9Yfhlx3w

- Kmeans Practical Problem II, www.youtube.com/watch?v=wObXFNsgcpw

- Kmeans Practical Problem III, www.youtube.com/watch?v=qPA7_zHGZ5I

- Kmeans Practical Problem IV, www.youtube.com/watch?v=fEMLHQ5Q484

- Kmeans Practical Problem V, www.youtube.com/watch?v=NSAa8IR4HBY

- Centrality measures and its applications

  https://en.wikipedia.org/wiki/Eigenvector_centrality

  https://en.wikipedia.org/wiki/Closeness_centrality

  https://en.wikipedia.org/wiki/Betweenness_centrality

CHAPTER 9

Case Study Part I

This chapter is the beginning part of the major consulting case study for this book. We will explain a typical Business Analytics consulting project and create a road map or an example of how to navigate a Business Analytics consulting project. We start with a description of the SME ecommerce environment generally, since this is the business environment of our selected case study; we then talk about the sources of data for analytics peculiar to this environment. Next, we describe the business to be used as a case study briefly, followed by the analytics road map peculiar to consulting for this business. This chapter ends with the results of the initial analysis and pre-engagement phase which forms the basis for the detailed analytics and implementation phase in Chapter 10.

9.1 SME Ecommerce

Ecommerce is the buying and selling of goods and services online. It involves trading of products or services directly or indirectly, by any means of venture, for the purpose of wealth creation, thereby adding value to the end users, and to improve means of livelihood.

Ecommerce small and medium enterprises (SMEs) are start-up ventures that employ digital online services as a platform for creating awareness and launching products and services to target customers. These involve promotional efforts and marketing strategies to initiate online purchases and drive sales and for maintaining repeated purchases.

© Afolabi Ibukun Tolulope 2022
A. I. Tolulope, *Data Science and Analytics for SMEs*,
https://doi.org/10.1007/978-1-4842-8670-8_9

The following are some of the analytics goals that one might want to achieve in an ecommerce analytics project:

- To capture, model, and analyze the behavioral patterns and profiles of users interacting with a website to recommend effective marketing campaign (i.e., extracts knowledge that may support several decisions, e.g., marketing campaigns per customer segment), pricing policy, etc.

- To understand customers to develop long-term relationships and to improve service quality.

- To discover consumer sentiments for effective customer relationship management.

- To discover communities of users who share common interests for better targeting.

- To provide recommendations for effective production plans, capital investments, and expansion decision of the company.

- To discover trends that can help to increase the popularity of the website among its visitors.

- To provide product cross-selling and layout recommendations to increase the number of products sold on the website and redesign the layout based on product recommendations.

- To optimize the sales funnel to make sure that as many people as possible go through the steps to buy the product, that is, to get more products sold on the website. In addition, discover weak spots in the sales funnel (ecommerce website) and recommend strategies to rectify them.

- To identify the most profitable customers, the most frequent search keywords, and the best sales timeline.

- To evaluate the website based on standard ecommerce key performance metrics (abandonment, micro- to macroconversion rates, average order value (AOV), sales conversion rates, customer retention rates, ecommerce churn rates, etc.) and use the results to provide recommendations to improve the sales on the website.

- To discover sequential or associative trends in customer buying patterns that could help to recommend for upselling, cross-selling, or embedding recommender systems.

- To develop a predictive model to determine which of the external customers (customers who have been to their website at least once and did not buy anything) have high propensity to buy so that we can recommend them for direct mailing or marketing.

- Customer lifetime value prediction and many more.

The types of data that one might analyze in such domain include

i. **Transaction data**

Transaction or business activity data has emerged rapidly as a result of interactions seen between consumers and the ecommerce business. These data are structured and come from a variety of sources, including customer relationship programs (e.g., customer profiles maintained by the company, the increase in the frequency of customer complaints) and sales transactions. Overall, it is clear that eretailers can reap numerous benefits from transaction data across the value chain.[3]

ii. **Clickstream data**

Clickstream data emanates from web and digital advertising, as well as social media activities such as ecommerce business tweets, blogs, Facebook wall postings, and many more. The series of pages viewed by a user within a specific web page is referred to as the clickstream, and this can be used for several types of analytics in ecommerce. The primary goal of clickstream tracking is to help get insight into the activities of visitors on the site.

iii. **Voice data**

Voice data is the information that is typically obtained through phone calls, call centers, or customer service. According to recent research, voice data can be useful for analyzing consumer purchasing behavior or targeting new customers.

iv. **Video data**

Video data is real-time data derived from current image capture. Ecommerce companies want to collect massive amounts of data in conjunction with image recognition tools, and not just swipe-stream information or transaction data.[4]

9.2 Introduction to SME Case Study

The case study business is a food and agriculture business named FarmCo. The focus of FarmCo is to make food items available at the doorstep and also supply farmers with the necessary inputs needed for their farming process. In essence, they help to connect food item suppliers and agricultural product suppliers with those that are demanding to purchase them. They take orders both offline and online, but even the offline orders are eventually entered into the online system so as to be able to have a holistic view to running the business operation. They have a staff strength of about 20 employees in different capacity; some are into the management of the offline operations such as logistics and supply, stocking and procurement, and so on, while others are into managing the online operations which include digital marketing and sales management. FarmCo is being managed by an ecommerce software which is the main data source for this analytics project.

The conceptual business model for FarmCo is given by Figure 9-1. It helps to understand the major parts and layout of the business process and to particularly know where the data for analytics will be coming from.

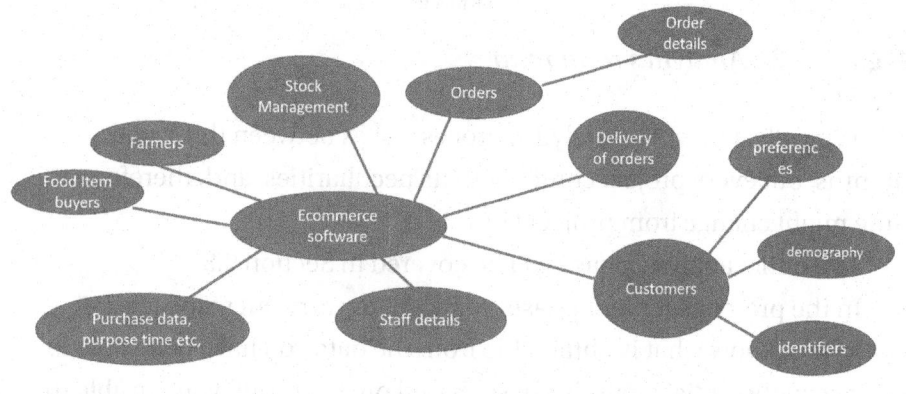

Figure 9-1. *Conceptual business model for FarmCo*

From Figure 9-1, we can see that the ecommerce software is at the heart of the operations process of the business; it helps to connect the farmers to those producing the equipment that they need and also connect the food item buyers to those that are producing them. The software also monitors the stock available, order, delivery, customer information, and some level of information of the staffs. The software is in sync with other software such as Google Analytics, etc. This is to be able to capture some information or data not built into the software.

For this project, the analytics road map is given in Figure 9-2.

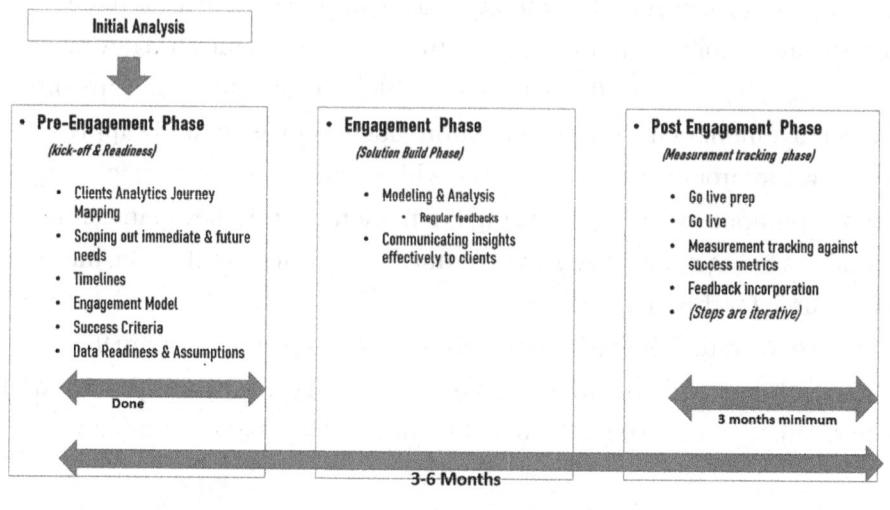

Figure 9-2. *Analytics road map*

Typically, the whole analytics process takes between three and six months, but every project comes with its peculiarities, and, therefore, this time might change from project to project.

The initial analysis phase will be covered in Section 9.3.

In the pre-engagement phase, you want to carry out some level of analysis to know what is obtainable from the data you have been able to retrieve in the initial analysis phase. Based on this result, you are able to present a detailed analytics journey mapping for the clients so that there will be a clear understanding of the roles of all parties in the project. At this point, you want to make it plain what is obtainable and what is to be moved into the future analytics project (recommendations to improve the analytics in the future). Timelines have to be specified with margins, and, most importantly, the stated goals each have to be clearly attached to the success criteria and how they will be measured. Finally, the data readiness (which translates to the feasibility of making the intended recommendations from the data) should be stated with the assumptions upon which the analytics project is to be carried.

In the engagement model, we develop our machine learning models and perform all necessary analysis with adequate communication (feedback) of the insights to necessary stakeholders.[1,2]

For the post-engagement phase, there is a need to train the teams responsible for the implementation of the recommendations from the analytics project. There is a need to explain how to successfully implement the result from the analysis. This is very important because it determines if the analytics project is termed successful or not. At the end of the day, even after an excellent work, if the analytics project insights are not correctly implemented, there might not be any or much success recorded. After the preparation, we now implement the results in whatever form applicable to the organization and give enough time to measure and track the metrics against the success criteria stated in the pre-engagement phase. Most times during or after the post-engagement phase, there might be a need to iterate that process based on feedback.

Figure 9-3 is the engagement model that clearly reveals all parties involved and what part they will be playing for the FarmCo analytics project. Figure 9-3 captures the engagement phase, where the data scientist reports to both the FarmCo stakeholders and the technical team during the modeling and analysis. Also the data scientist during modeling receives feedback from the technical team for necessary model reworking. In the post-engagement phase, both the data scientist and the technical team are involved in the measurement and tracking of the goals, metrics, and success criteria after successful implementation is carried out. Note that in Figure 9-3, feedback also goes to the implementation stage, and the cycle might be iterated based on results obtained from the modeling and measurement phases.

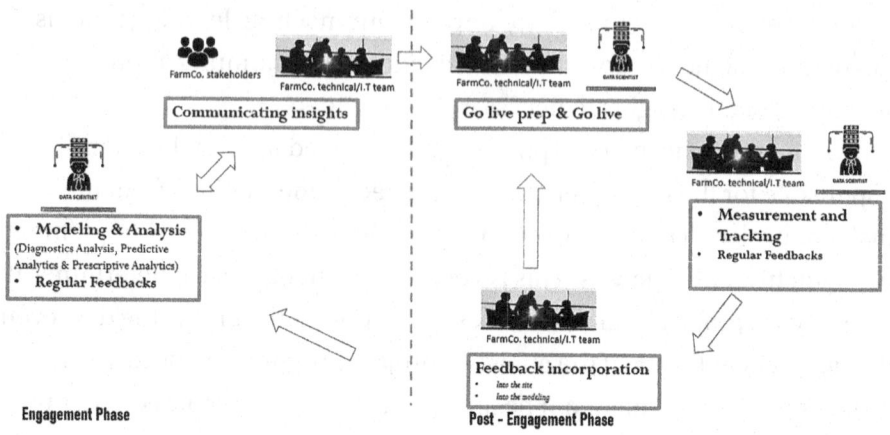

Figure 9-3. *Engagement model for the FarmCo analytics project*

9.3 Initial Analysis

The initial analysis has to do with the following steps:

1. Highlight the decision objectives (goals) and determine the analytics output that will help to make intended decision feasible.

2. Discover available data (internal and external) for analytics goals in (1).

3. Access the data in (2) and what is the quality for analytics goals in (1).

4. Is the data in (3) able to determine the analytics output in (1)?

During the initial analysis for FarmCo, the goals that the stakeholders intended to solve (or they proposed) include the following:

1. Reach out to prospects and convert them to loyal customers.

2. Maximize social media engagement, that is, tracking and converting the engagement to sales.

3. Provide recommendations for effective production plans, capital investments, and expansion decision of the company.

4. Discover trends that can help to increase the popularity of the website among its visitors.

5. Identify best-selling items.

Based on the result of steps 1 to 4 of the initial assessment, the goals were modified to the following after several feedback meetings with the stakeholders:

1. Increase website sales revenue.

2. Increase and retain (customer satisfaction) website traffic.

Note that there were several recommendations for data platform upgrades that will help to explore some of the intended goals directly, for example, to reach out to prospects and convert to loyal customers, there has to be a system or program module that collects some information from the prospects so that they can be the object of targeting after a segmentation modeling.

9.4 Analytics Approach

For each of the goals to be explored in this analytics project, we will use the analytics approach described in Section 1.3. Just as mentioned earlier, depending on the application situation, we might first use descriptive analytics as a form of exploratory research to discover questions to be answered in the diagnosis analysis stage and then follow this up with predictive and prescriptive. Other times, we might just go straight into any of them. All these decisions are determined by the nature of the analytics project.

Goal 1: Increase website traffic

Using descriptive analytics, we need to find out the current state of the website traffic. We want to answer questions like the following:

1. *Is the website traffic increasing or decreasing?*

   - Why is the traffic going up and down? Why is it not increasing steadily?

   - What special thing did they do in the months of March and May to give a slight jump in visitors?

   - Anything happened in July to reduce the website traffic

   - Could it be a decrease in published content or loss of backlinks, etc.?

   - Could it be server failure, holidays, etc.?

   - Where are the traffic coming from? Can the visits be broken down to the sources of the traffic?

   - Break the traffic down to new users and returning customers, and where are they coming from?

- What marketing techniques have they been using? Which ones have been successful and which ones have not? It will be nice to use data to answer these questions.

Goal 2: Increase website sales revenue

The first thing to do is to use descriptive analytics to find out the current situation of things as regards this goal. Particularly, we want to explore the following metrics sales, revenue, and conversion rate. These are the metrics that can help to measure and understand this problem or goal. We want to answer questions like the following:

Are the sales and revenue of the website incremental, fluctuating, plateauing, or declining?

The result of the descriptive analytics is presented in Section 9.5 for the two goals of this analytics project. For goal one, the result of the descriptive analytics shows that the revenue is fluctuating contrary to the ideal which should at least be increasing steadily. This then required that there be a follow-up diagnosis analytics to find out why this is so. In the diagnosis analytics, we want to answer the following questions:

Which marketing activities increased orders?

What is the effectiveness of marketing campaigns as per order?

Where are the orders coming from?

What categories of products are bringing in the order?

What is the AOV (average order value), which tracks the average dollar amount spent each time a customer places an order on a website?

What marketing activities increased revenue?

What is the effectiveness of marketing campaigns as per revenue?

Is the revenue incremental?

What is determining the conversion rates?

- *What paths are customers taking to check out?*

- *What is responsible for low conversion rate?*

- *Why are visitors not buying or completing the checkout process?*

- *What is their checkout process like? Is it complicated?*

The answers to these questions will be revealed in the pre-engagement section (Section 9.5). These answers, together with the available data, will then determine the predictive modeling approach that is described in Section 10.1.

9.5 Pre-engagement

In this section, we will give examples of some of the descriptive analytics outputs for answering some of the questions asked in the previous section. Also, in the pre-engagement phase, there is a need to detail the metrics to be used in assessing the success of these goals. Also, at this stage after getting the results from the descriptive analytics, we need to decide if recommendations will be made for implementation (i.e., we have been able to discover some insights that the business can use to make some decisions) or for further analysis (i.e., the results will form the basis for the modeling in the engagement phase) or for future analytics (i.e., there is a need to provide data advice that could help the analytics in the future).

Goal 1: Increase website traffic

For this goal, the success criterion is the website traffic, and the current situation of such is depicted in Figure 9-4.

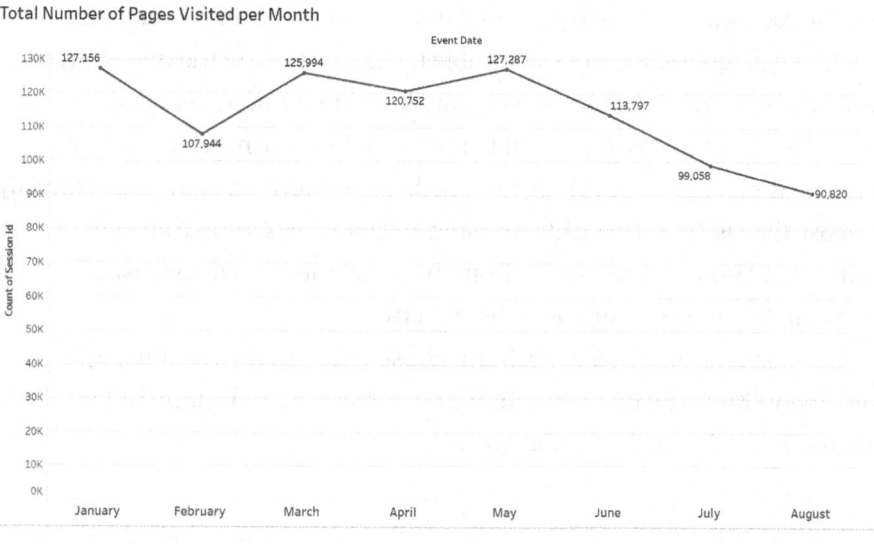

Total Number of Pages Visited per Month

Figure 9-4. *Current website traffic*

From Figure 9-4, we can see that the traffic goes up and down, with a steady decline from June. At this point, we want to find out if there is any reason for this from the organization's point of view, for example, change in marketing, weather, government issues, and so on. Ideally, the traffic is supposed to be increasing steadily.

There are several angles to these descriptive analytics that could be presented based on the available data, for example, we can also try to see the top ten most visited pages and the time of these visits. In addition, for each marketing channel, SEO, email campaign, coupon code, social media, etc., report the new visitors, returning visitors, orders, sales, conversions, new online accounts, and bounce that has come from them in total (i.e., not monthly). The result of this will help to know which channel is bringing in the visits, for example, and the diagnosis can start from there. We will also be able to know if the investments in these channels are proportional to the visit's results.

The recommendation here is that there needs to be further analysis preferably diagnosis analytics to be able to find answers to the questions that have come up. In this book, we will provide examples of such diagnosis report in Section 10.1, but it is important to note that there are much more that could be done in this regard not included. In addition, we plan to also directly increase the visit to the website by using advanced descriptive analytics to recommend how to direct traffic from the social media to the website.

Goal 2: Increase website sales revenue

For this goal, the success criteria chosen are orders, revenue, and conversion rate. The current situation of these criteria is depicted in Figures 9-5, 9-6, and 9-7, respectively.

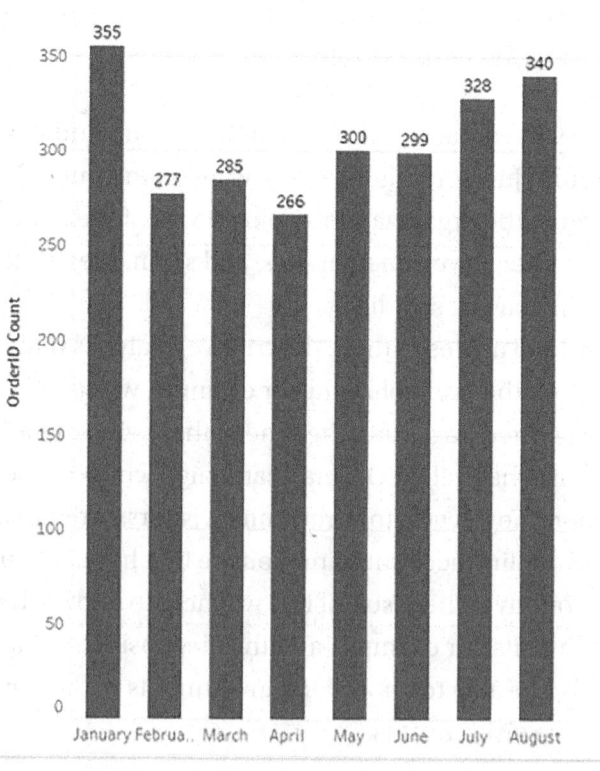

Figure 9-5. *Current state of orders from the website*

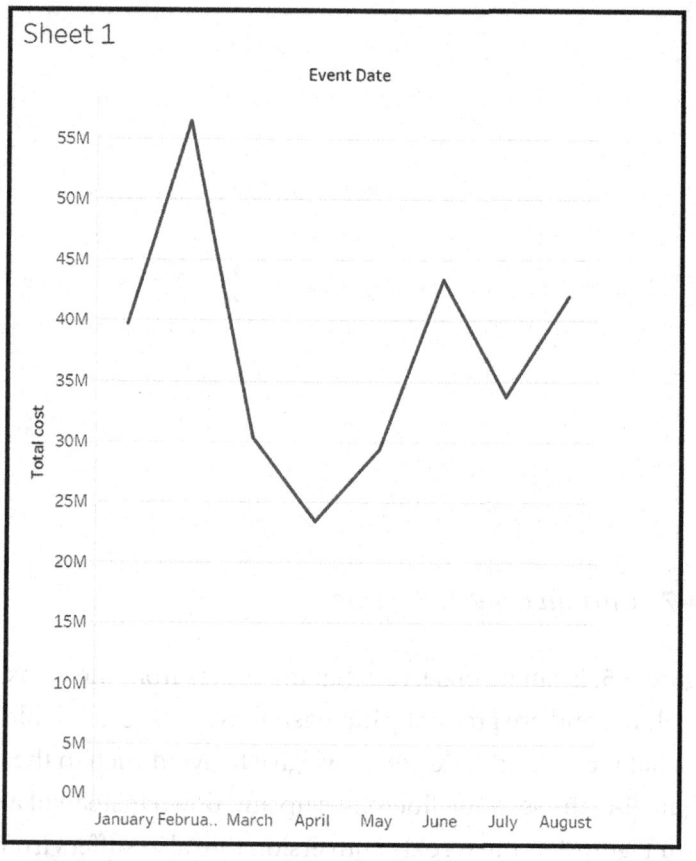

Figure 9-6. *Current situation of the revenue*

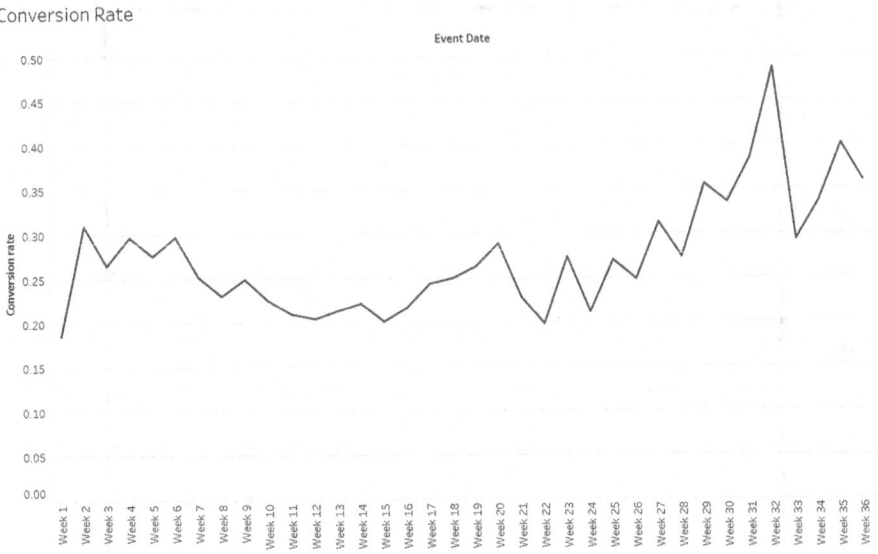

Figure 9-7. *Current conversion rate*

In Figure 9-5, it can be observed that the orders from the website slightly declined and are gradually increasing. What is responsible for this decline? What recommendations can we give to avoid such in the future?

In Figure 9-6, the revenue fluctuates up and down instead of a steady increase. In Figure 9-7, the weekly conversion rate also suffers from the same fluctuation trend.

Apart from the recommendation for the further diagnosis analytics for this goal, we would also recommend some modeling in the predictive and prescriptive approaches in order to directly increase the sales and revenue from the website. This is also covered in Section 10.1.

9.6 References

1. Devin Pickell (February 28, 2019) 4 Types of Data
    Analytics Your Business Can Benefit. From `www.`
    `g2.com/articles/types-of-data-analytics`

2. Valerie Lavskaya (September 10, 2019) THE
    THREE-LEVEL ANALYTICS APPROACH FOR
    ECOMMERCE: DESCRIPTIVE, PREDICTIVE
    AND PRESCRIPTIVE, `www.promodo.com/blog/`
    `the-three-level-analytics-approach-for-`
    `ecommerce-descriptive-predictive-and-`
    `prescriptive/`

3. Akter, S., & Fosso Wamba, S. (2016). Big data
    analytics in E-commerce: A systematic review and
    agenda for future research. *Electronic Markets, 26*, 0.
    `https://doi.org/10.1007/s12525-016-0219-0`

4. Singh, R., Verma, N., & Gupta, M. (2020).
    ROLE OF E-COMMERCE IN BIG DATA. 11(12),
    1770–1777. `https://doi.org/10.34218/`
    `IJARET.11.12.2020.166`

CHAPTER 10

Case Study Part II

In this chapter, we will conclude the case study used for the illustration of a typical Business Analytics consulting for an SME by presenting the details of the engagement phase for the case study in question. The post-engagement phase is left out as the implementation of the recommendations is determined by the systems and procedures of the business. It is important to note that the consulting steps can be customized for any business (particularly small businesses). The whole steps described in Chapters 9 and 10 have been made simple for understanding, though in real-life business applications, there might be a need to iterate the process until satisfactory results have been obtained. This is because you constantly need to incorporate feedback from the stakeholders and domain experts.

10.1 Goal 1: Increase Website Traffic

The first thing is to state clearly the analytics procedure concluded from the pre-engagement phase for the first goal. To address the first goal of increasing website traffic, we will use two major analytics approaches given the available data.

1. From the result of the pre-engagement phase, we will follow up with a diagnosis analysis to discover what is responsible for the nature of the current website traffic and make recommendations from

A. I. Tolulope, *Data Science and Analytics for SMEs*,
https://doi.org/10.1007/978-1-4842-8670-8_10

the results. Note that in this section, only the results of the diagnosis analysis are presented. For this, we will use descriptive analytics.

2. We will study the social media environment in which FarmCo operates and attempt to identify some major influencers in that environment for the purpose of marketing and directing traffic. For this, we will use social network analysis.

Analytics result

1. Several data sources were combined to address this goal, data from marketing, sales promotion, and also data from the Google Analytics account of FarmCo. The data was used to answer the following questions using descriptive analytics.

2. *Why is the traffic going up and down?*

From the interview with the sales and marketing team, it was discovered that the reduction in the month of May and June is based on the reason that there was no advert on Facebook, etc., but FarmCo relied only on organic search. The increase in traffic experienced in the month of March is due to a promotion in the month of February middle promotion.

3. *What are the inputs to the website traffic?*

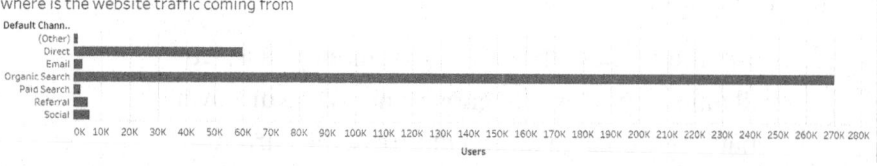

Figure 10-1. *Inputs to the website traffic*

The result in Figure 10-1 is used to evaluate the budget available for these platforms and see if it is directly proportional to the traffic results obtained from them.

4. *What kind of traffic are they getting?*

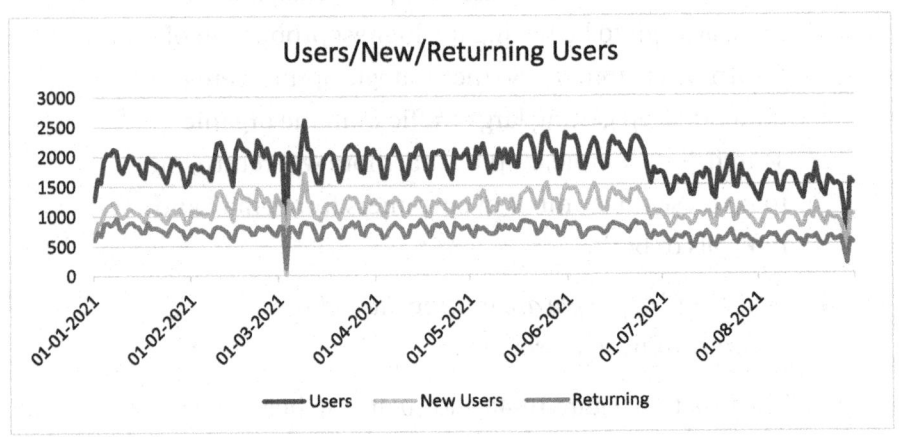

Figure 10-2. *Type of traffic*

The result of the type of traffic (Figure 10-2) will help them to know who is actually interested in visiting their website; is it new or returning visitors? This information can be used in several ways; an example is to know how to modify their target audience when marketing.

5. *Where is the traffic of new customers/returning customers coming from?*

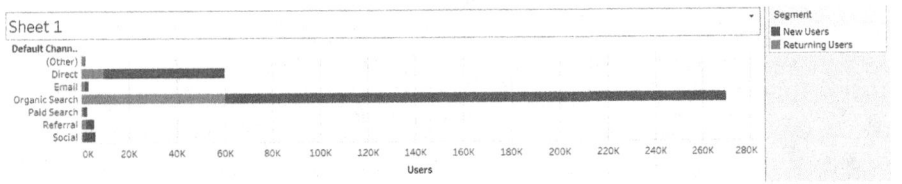

Figure 10-3. *Source of traffic for new/returning customers*

From the domain expert, it was gathered that returning customers are more likely to do business, and this makes us particularly interested in the returning customers. From Figure 10-3, we can see that though small in number, the email and referral seem to be having the highest proportion of returning customers. So the strategic option here is either to keep getting large traffic from the organic search, for example, hoping that more will return, or to increase the number of customers from email and referrals or both.

6.   *What kind of people are visiting the website (demography, interests, etc.)?*

The results in Figure 10-4a and 10-4b is to help evaluate if the interest and demography of the target audience used in marketing tallies with that of those actually coming to the website? Conclusions from this will help to adjust the marketing audience targeted.

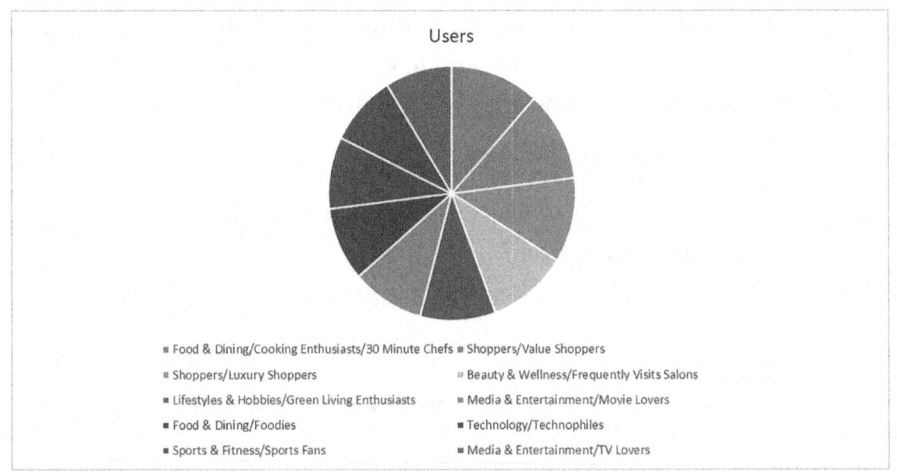

Figure 10-4a. *Demography of visitors (interests)*

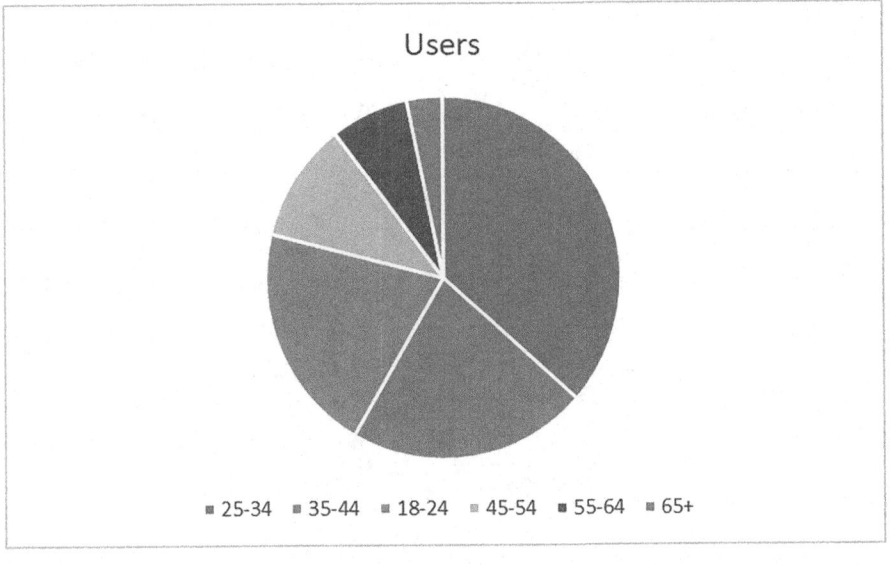

Figure 10-4b. *Demography of visitors (age)*

In summary, there is a need to take decisive steps on the timing of promotional activities. There is also a need to adjust the target audience for marketing to increase traffic. Also discovered are the valuable marketing channels/campaigns to focus the budget on to increase the traffic.

7. For this demonstration, we pick only one product of FarmCo, which is poultry; for this product, we use the search criteria (poultry, Nigeria) to extract tweets on netlytic.org. We are able to create the communication network of people interested in this product in Nigeria. The data to be used for this practical demo is named PoultryNigeria.gephi. After loading the data, running all the statistics, and performing necessary cosmetics for the network (eigenvector centrality was used), the resulting network is given in Figure 10-5.

287

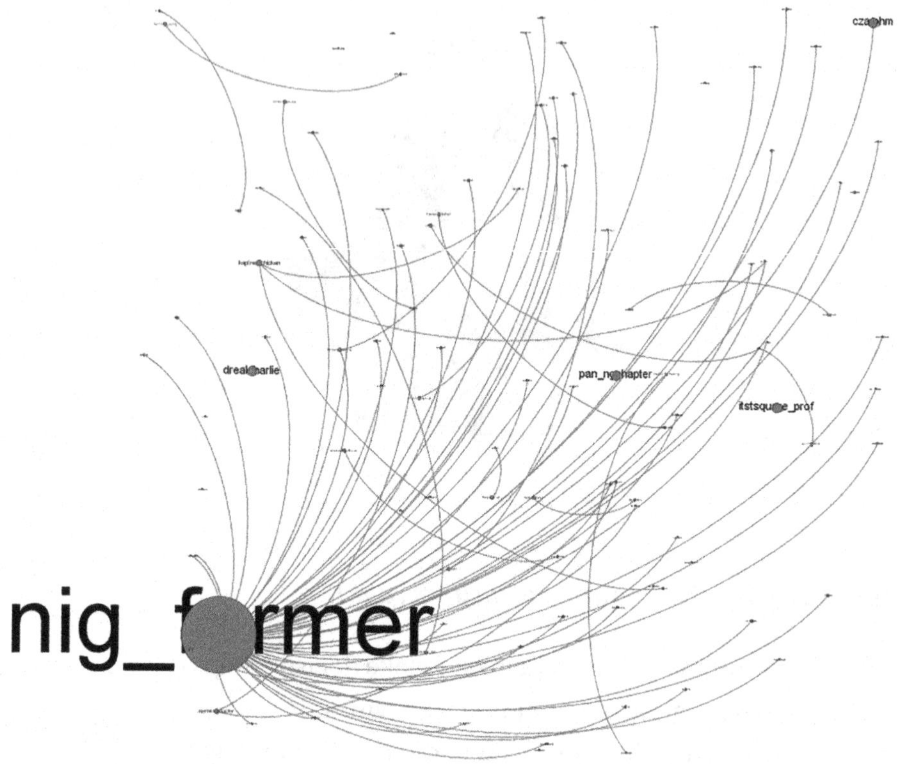

Figure 10-5. *Visualized Network of FarmCo*

From Figure 10-5, the username nig_farmer is the major influencer and can be used for reaching out for marketing purposes.

10.2 Goal 2: Increase Website Sales Revenue

Just like the first goal, we first state clearly the analytics procedure concluded from the pre-engagement phase for the second goal. To address the second goal, we will use three major analytics approaches given the available data:

1. *Customer loyalty intervention*: Apply logistic regression on sampled historical data to build a model to discover the customers that will churn. This model will be used to predict what the existing customers on their website will do (churn or not churn). This will then be followed up with associative rule mining, which will be used to recommend products to those that would have churned through email. In this approach, churn is described as those that have not bought anything in the last two years back from the last date that the historical data was sampled.

2. *Recency, frequency, and monetary segmentation*: For the second approach, we will perform RFM segmentation. This was a decision based on the lack of attribute richness for the typical clustering. The intention is to be able to understand and profile the current customers of FarmCo for better understanding toward customer relationship management.

3. *Strategic market target*: This approach will model prescriptive analytics. Prescriptive analytics will be used to determine the best actionable strategy for targeting future customers. For this analytics approach, the clustering technique will be used, k-means clustering to be specific. We will also be adopting the thinking backward approach in the analytics project.

Customer loyalty intervention

1. ***Import*** the data named *SampleHistorical.xlsx*, replace errors with missing values, and exclude CustomerID, Billing State, Email Permission, Device Brand, Shipping Country, and Shipping State (this is due to either too many categories or only one category).

2. Check the statistics to be sure that there are no missing values.

3. Visualize the attribute for the classification task using the following as a guide:

   a. Study the target variable to categorical outcome using a bar chart, outcome on y axis (example in Figure 10-6). Doing this for all relevant attributes revealed that (apart from Email Status, Os Version, and Payment Method) the categories of all the categorical attributes are fairly OK as there are no attributes with too many categories or error in the categories or imbalanced categories. If any of these issues existed, it needs to be dealt with at this point.

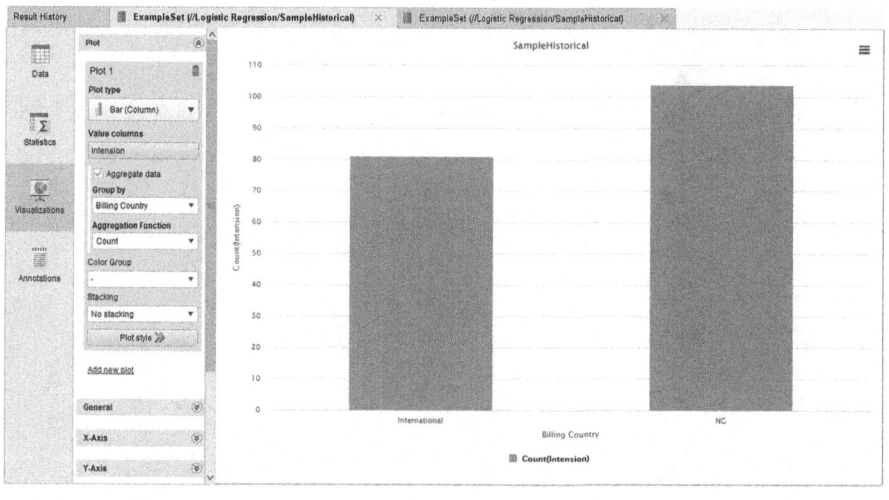

Figure 10-6. *Example of studying the target variable to categorical outcome*

    b.   Study the relationship of the target/outcome variable to pairs of numerical predictors via a color-coded scatter plot (example in Figure 10-7). Doing this for all relevant variables revealed that there is no correlation between the numerical predictors. This therefore means that we can move all the numerical predictors to the modeling stage.

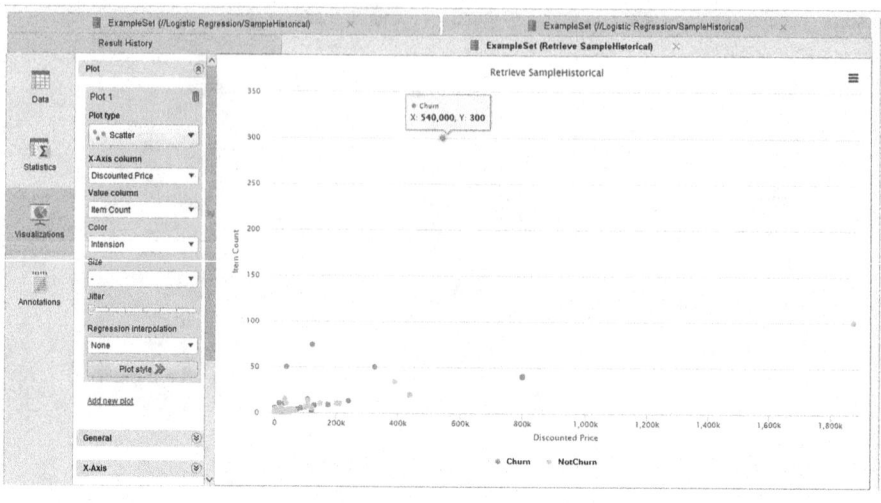

Figure 10-7. *An example of studying the relationship of the outcome variable to pairs of numerical predictors via a color-coded scatter plot*

    c.   Use correlation analysis to remove correlated attributes (Figure 10-8). The results of the correlation revealed that there is a need to remove either Item Count or Quantity as they are highly correlated.

Attribut...	Billing C...	Browse...	Device ...	Email St...	Intension	Os (gro...	Os Vers...	Paymen...	Discoun...	Item Co...	Quantity	Total A...	Unit Price
Billing C...	1	?	?	-0.080	0.019	?	?	?	0.109	-0.066	-0.042	0.069	0.155
Browser ...	?	1	?	?	?	?	?	?	?	?	?	?	?
Device T...	?	?	1	?	?	?	?	?	?	?	?	?	?
Email St...	-0.080	?	?	1	-0.031	?	?	?	-0.062	-0.068	-0.069	0.035	0.050
Intension	0.019	?	?	-0.031	1	?	?	?	0.100	-0.010	-0.001	0.210	0.110
Os (group)	?	?	?	?	?	1	?	?	?	?	?	?	?
Os Versi...	?	?	?	?	?	?	1	?	?	?	?	?	?
Payment...	?	?	?	?	?	?	?	1	?	?	?	?	?
Discount...	0.109	?	?	-0.062	0.100	?	?	?	1	0.612	0.621	-0.076	-0.005
Item Cou...	-0.066	?	?	-0.068	-0.010	?	?	?	0.612	1	0.989	-0.060	-0.202
Quantity	-0.042	?	?	-0.069	-0.001	?	?	?	0.621	0.989	1	-0.054	-0.173
Total Am...	0.069	?	?	0.035	0.210	?	?	?	-0.076	-0.060	-0.054	1	-0.009
Unit Price	0.155	?	?	0.050	0.110	?	?	?	-0.005	-0.202	-0.173	-0.009	1

Figure 10-8. *Correlation matrix*

4. Using the conclusions from step 3a–c, you can create a logistic regression model as seen in Figure 10-9. (Note that it is better to use stratified cross-validation as the data sample is small.) The evaluation of the model is seen in Figure 10-10. The model gave an accuracy of 61.08%. (Note that at this stage, there is a need to try other algorithms, attribute selection techniques, and so on, if we desire a higher accuracy, but for the purpose of this illustration, we proceed regardless.)

Warning: Removed collinear columns [Billing Country = NG, Browser (group) = Chrome Mobile, Device Type (group) = M, Os (group) = Android]

Attribute	Coefficient	Std. Coefficient	Std. Error	z-Value	p-Value
Billing Country = International	0.709	0.353	0.382	1.856	0.064
Billing Country = NG	0	0	?	?	?
Browser (group) = Chrome	-0.145	-0.066	1.136	-0.128	0.899
Browser (group) = others	4.562	2.031	4.096	1.114	0.265
Browser (group) = Small Browser	0.208	0.072	0.778	0.267	0.790
Browser (group) = Chrome Mob...	0	0	?	?	?
Device Type (group) = D	0.424	0.201	1.789	0.237	0.813
Device Type (group) = Others	-2.082	-0.944	3.566	-0.584	0.559
Device Type (group) = M	0	0	?	?	?
Os (group) = Windows	-1.487	-0.685	1.839	-0.809	0.419
Os (group) = Others	-2.475	-1.176	1.395	-1.774	0.076
Os (group) = Android	0	0	?	?	?
Discounted Price	0.000	0.433	0.000	1.374	0.170
Quantity	-0.008	-0.218	0.016	-0.514	0.607
Total Amount	0.000	0.857	0.000	2.611	0.009
Unit Price	0.000	0.184	0.000	1.075	0.282
Intercept	-0.979	-0.413	0.410	-2.388	0.017

Figure 10-9. *Logistic regression model*

accuracy: 61.29% +/- 12.94% (micro average: 61.08%)

	true Churn	true NotChurn	class precision
pred. Churn	87	49	63.97%
pred. NotChurn	23	26	53.06%
class recall	79.09%	34.67%	

Figure 10-10. *Evaluation of the logistic regression model*

5. Using the modeling in Figure 10-9, we can predict the intention of the current customers to churn using the data named *FoodAgricGroup.xlsx*. Take note that this data does not contain the attribute named Intention, as this is what is to be predicted. The result of the prediction can be seen in Figure 10-11.

Row No.	prediction(In...	confidence(Churn)	confidence(NotChurn)	Billing Count...	Browser (gr...	Device Type ...	Email Status	Os (group)	Payment Me...	Discou
1	Churn	0.670	0.330	NG	Chrome Mobile	M	A	Android	BankTransfer	13250
2	Churn	1.000	0.000	International	Chrome Mobile	M	A	Android	BankTransfer	42300
3	Churn	1.000	0.000	International	Others	M	A	Android	BankTransfer	66900
4	Churn	1.000	0.000	International	Chrome	D	A	Windows	BankTransfer	500
5	Churn	1.000	0.000	International	Chrome Mobile	M	A	Android	BankTransfer	11300
6	Churn	1.000	0.000	International	Others	Others	A	Others	BankTransfer	87500
7	Churn	0.988	0.012	International	Others	Others	A	Others	DebitCreditC...	33500
8	Churn	0.763	0.237	International	Chrome	D	A	Windows	BankTransfer	12200
9	Churn	0.933	0.067	International	Others	M	A	Others	BankTransfer	49900
10	Churn	0.895	0.105	International	Chrome	D	A	Windows	BankTransfer	27800
11	Churn	0.777	0.223	International	Chrome	D	A	Windows	BankTransfer	24900
12	Churn	0.739	0.261	International	Chrome Mobile	M	A	Android	BankTransfer	59500
13	Churn	0.982	0.018	NG	Others	Others	A	Others	BankTransfer	30000
14	Churn	0.876	0.124	NG	Others	Others	A	Others	BankTransfer	90000
15	Churn	0.884	0.116	NG	Others	Others	A	Others	BankTransfer	10000
16	Churn	0.993	0.007	NG	Others	Others	A	Others	BankTransfer	40000
17	Churn	0.995	0.005	International	Chrome	D	A	Windows	DebitCreditC...	22150
18	Churn	0.870	0.130	International	Chrome Mobile	M	A	Android	BankTransfer	17230

ExampleSet (802 examples, 3 special attributes, 11 regular attributes)

Figure 10-11. *Prediction Result*

6. The next step is to use the transactional data named *FoodAgricTransAsso.xlsx* to perform the association rule to discover items that are being sold together. The result of this is revealed in Figure 10-12.

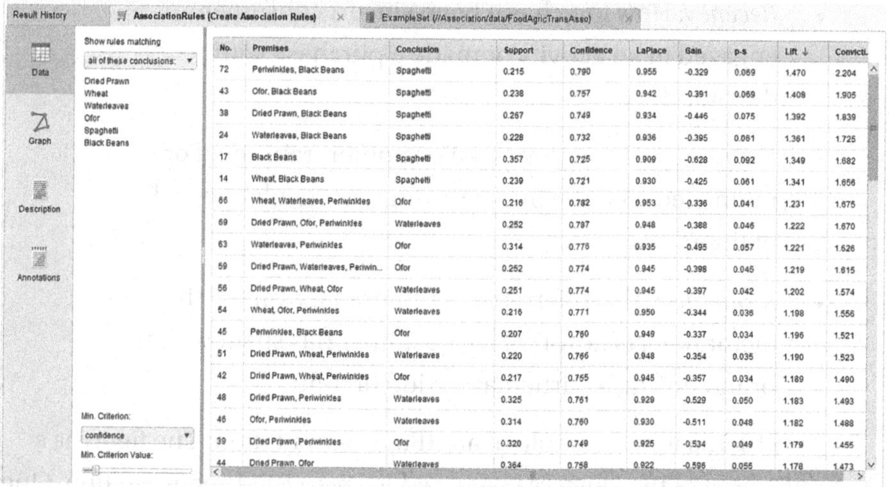

Figure 10-12. *Result of association rule mining*

7. To conclude this analytics approach, we will use the inference in Figure 10-12 to recommend products to customers (with high propensity to churn as revealed in Figure 10-11) through an email marketing campaign.[2,3] Note that this is just one approach to implementing the analytics results. There are other creative ways that can be explored such as embedding a recommendation engine on the website and many more.

RFM segmentation

RFM segmentation[1] is used to target certain clusters of clients with communications that are more tailored to their specific behaviors. RFM segmentation is a useful tool for identifying groups of clients who should be given extra attention. The letters RFM stand for recency, frequency, and monetary value. RFM uses numerical scales to produce a high-level portrayal of clients that is both brief and instructive. It is straightforward to comprehend and interpret.

- *Recency*: How long has it been since a consumer engaged in an activity or made a purchase with the brand?

- *Frequency*: How often has a customer transacted or interacted with the brand during a particular period of time?

- *Monetary*: Also referred to as "monetary value," this factor reflects how much a customer has spent with the brand during a particular period of time.

Figure 10-13a is an example of an RFM segmentation. The figure is a 3D visualization of the clusters generated from an RFM segmentation. One of the interpretations from Figure 10-13a reveals that some customers have not spent a lot of money but frequently visit the business website and have made recent purchases of high value.

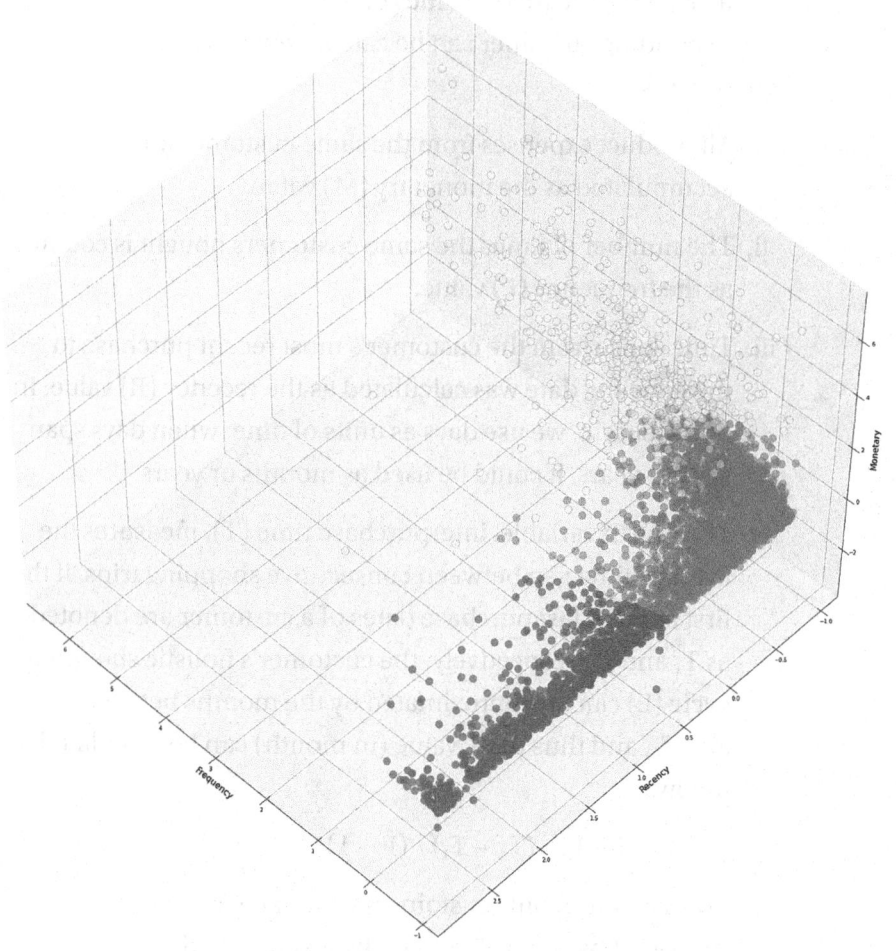

Figure 10-13a. *3D visualization of RFM segmentation[4] (https://*
miro.medium.com/max/1400/1*_uQS11qVzIv4rTNDPM6Rqw.png)

To perform customer segmentation using the RFMT (Recency
Frequency Monetary Tenure) approach, the spreadsheet data must be
transformed to RFMT data. The following tips can help:

    a.   Create a unique identification for each customer.

b. For a given ID, the RFMT values of the corresponding customer can be calculated from the other fields.

    i. All product expenses from the same customer are accumulated as the monetary (M) value.

    ii. The number of times the same customers bought is counted as the frequency (F) value.

    iii. Time-lapse from the customer's most recent purchase to the crawling date was calculated as the recency (R) value. In this scenario, we use days as units of time; when days span several years, it could be used as months or years.

    iv. The fourth variable, interpurchase time (T), measures the average time gap between consecutive shopping trips. If the first and the last purchase dates of a customer are denoted as T_1 and T_n, respectively, the customer's holistic shopping cycle (L) can be approximated by the months between T_1 and T_n, and thus the T value (in month) can be calculated as follows:

$$T = L / (F - 1) = (T_n - T_1) / (F - 1) \ (1).$$

To calculate T, only customers who made at least two purchases (i.e., $F \geq 2$) in the given period were considered.

This resulted in the data named *RFMFoodAgric.xlsx*.

c. The RFMT values obtained in b are then used to divide the customers into four tiers for each dimension, such that each customer will be assigned to one tier in each dimension as seen in Table 10-1.

Table 10-1. *Four Tiers for Each Dimension*

Recency	Frequency	Monetary
R-Tier-1 (most recent)	F-Tier-1 (most frequent)	M-Tier-1 (highest spend)
R-Tier-2	F-Tier-2	M-Tier-2
R-Tier-3	F-Tier-3	M-Tier-3
R-Tier-4 (least recent)	F-Tier-4 (only one transaction)	M-Tier-4 (lowest spend)

This resulted into the data named *FinalRFM.xlsx* (though this data has about five tiers instead of four as in Table 10-1).

It is possible to use the results of RFM segmentation to target certain groups of clients depending on the RFM segments in which they appear. Groups like best customers are examples of this type of consumer (customers who are found in R-Tier-1, F-Tier-1, and M-Tier-1, meaning that they transacted recently, do so often, and spend more than other customers).

High-spending new customers: Customers in 1-4-1 and 1-4-2. These are consumers that have only transacted once, but very recently, and have spent a significant amount of money.

Lowest-spending active loyal customers: Customers in segments 1-1-3 and 1-1-4 make up this group. They transacted recently and do so often, but spend the least.

Churned best customers: Customers in groups 4-1-1, 4-1-2, 4-2-1, and 4-2-2. They used to transact frequently and spend a lot of money, but they haven't done so in a long time, and so on.

1. Import the data named *FinalRFM.xlsx*, check the statistics, and make sure there is no missing data. Change the type of all the RFMT attributes to polynominal.

2.   Visualize the data using a bar chart to see that the RFMT categories have fair data distribution. Figure 10-13b is an example.

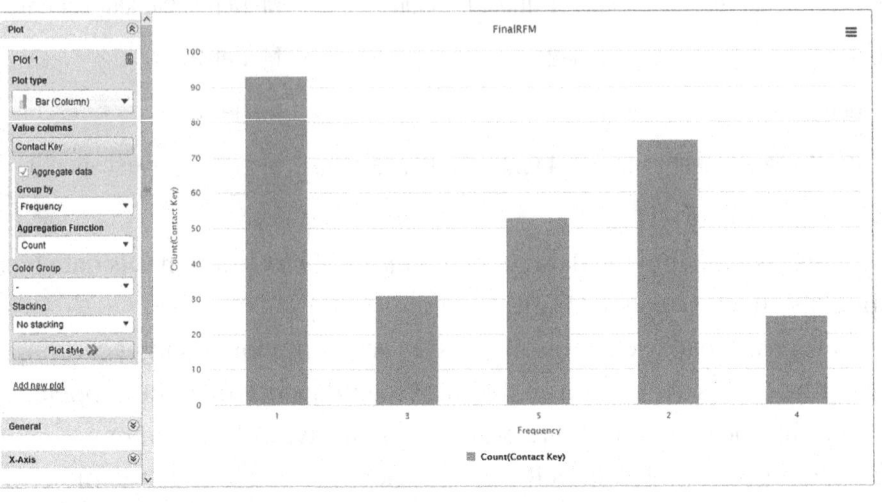

Figure 10-13b. *Bar chart for the frequency*

3.   Figure 10-14 is the final cluster visualization for the RFMT modeling, when k is equal to 4, for example.

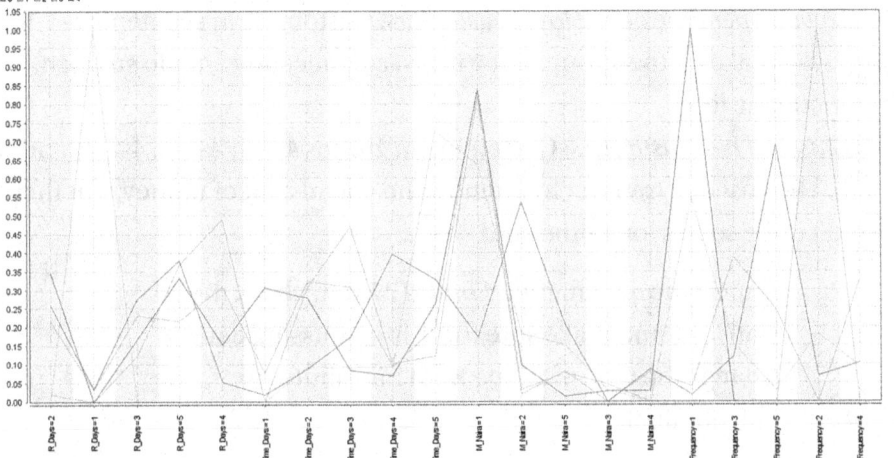

Figure 10-14. *Clusters discovered*

4. Based on the clusters discovered and the tiers used
    for the RFMT data, we can profile the clusters. The
    final step in RFM segmentation is to customize
    communications for each of the discovered clusters
    of customers.

Strategic market target

The goal here is to prescribe the optimal cluster in the external data to
target for marketing using prescriptive analytics. We will use the thinking
backward approach for this assignment. This approach consists of the
following steps:

- Start with the decision we want to make.

- Determine what analysis outputs would help make that
  decision.

- Design the analysis that creates those outputs.

- Determine what data is needed for analysis and how
  to get it.

- Execute the analysis.

- Display and present the result.

1. *Start with the decision we want to make*: In this
    project, we need to choose an actionable customer
    acquisition strategy out of the following available
    options.

    a. Increase customer base by acquiring as many customers as
       possible, no matter what kind of customers they are, with the
       hope of retaining them for high future revenue.

    b. Increase customer base by acquiring only few premium
       customers (due to low marketing budget).

    c. Increase customer base by acquiring many high profit customers (best-case scenario).

2. *Determine what analysis outputs would help make that decision*: The analysis output we are expecting is to have one or two slides that describe each option in step 1 and the summary that compares each option based on key metrics (available in the data). We will also explain how we arrived at the conclusions. Note that this procedure leverages on having the same attributes in internal and external data. In some situations, this might not be possible, but there are third parties that can help with external segmentation to overcome this.

3. *Design the analysis that creates those outputs*: Figure 10-15 is a high-level conceptual design of the approach. The aim is to segment the internal data and also segment the external data. After that, we will now link the two together and construct options which will help to maximize the business decisions in step 1. The intention is to use internal data to describe the characteristics of the current customer base and the historical performance of these segments (e.g., total profit made from the customers). We will follow this up by describing the external customer characteristics and use these to determine what the acquisition opportunity is for each segment.

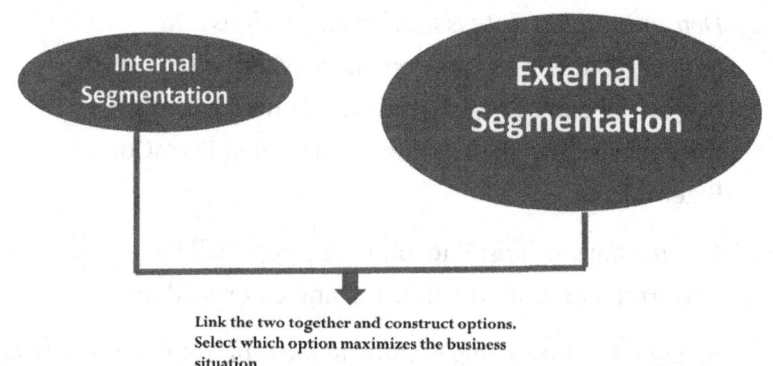

Figure 10-15. *High-level conceptual design*

In Figure 10-16, we have a more detailed version of the conceptual schema revealing the proposed clusters in the internal data and the external data. (This is just to give an idea of what we are expecting as in practice the knowledge of the clusters might not be clear like this in the beginning.)

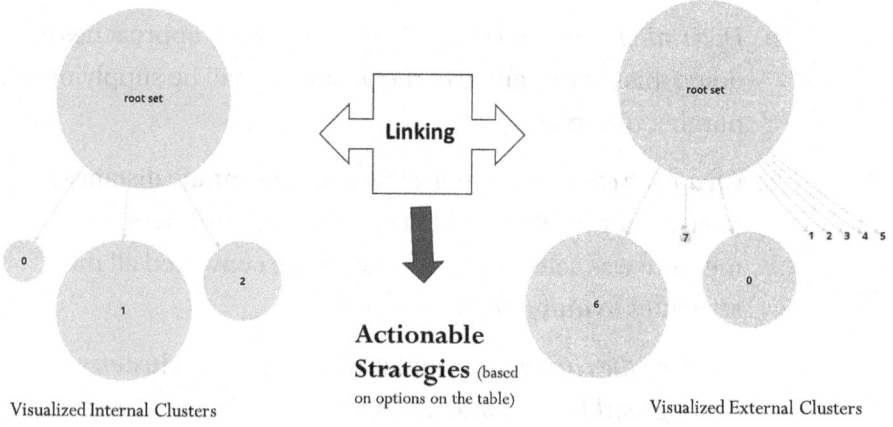

Figure 10-16. *More detailed conceptual design*

4. *Determine what data is needed for analysis and how to get it*: For this approach, we will need to use two sets of data. The internal data named FarmCo. Internal.xlsx and the external data named FarmCo. External.csv.

5. *Execute the analysis*: The following steps will be performed for both the internal and external data.

    a. *Data preprocessing*: *Missing value handling* - k-means cannot deal with missing values. If there are only few missing values, they can be excluded; if there are many, they have to be imputed. *Data visualization* will be used to discover outliers. Convert *categorical variables to numerical as* k-means uses a distance function and cannot handle categorical variables directly (if it's ordinal, replace with an arithmetic sequence; if it's nominal, convert to binary). *Data normalization will be used* to weigh each dimension equally.

    b. *Determine the number of k*: There are several approaches to doing this (as explained earlier), but we shall be supplying the number of k to use.

    c. *Create the clustering model*: Use the appropriate distance measure for the distance between records (numerical measure was selected because we have converted all the attributes to numerical).

    d. *Validate the clusters*: To validate the generated cluster, we use the *avg. within-centroid distance*.

    e. *Interpret your results*: Compare the cluster centroid to characterize the different clusters, and try to give each cluster a label.

Internal data 5a–e

After examining the internal data for missing values, we have the statistics (scroll down to see all) in Figure 10-17.

			Min	Max	Average
✓ Customer ID	Integer	0	1	21	11
✓ No of C	Integer	0	0	15	4.381
✓ No of D	Integer	0	7	44	20.952
✓ No of N	Integer	0	4	82	25.524
✓ Quantity	Integer	0	0	6	2.667
✓ Item Count	Integer	0	4	61	31.476
✓ Total Amount	Integer	0	120	440	280
✓ No of Days	Integer	0	0	14	2.286
✓ Discounted Price	Integer	0	0	69	27.429
			Min	Max	Average

Figure 10-17. *Summary statistics*

Checking the numerical attributes for outliers (using scatter plots), we discover that all the attributes don't have outliers. Figure 10-18 is an example.

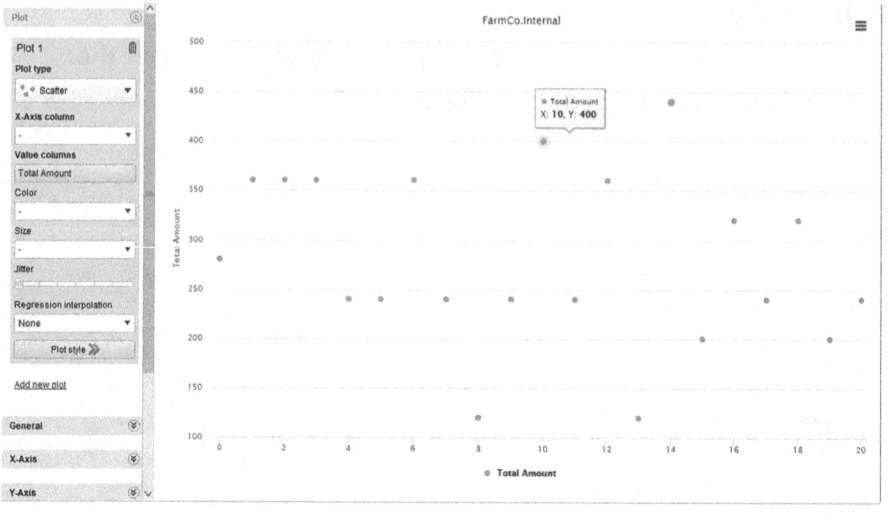

Figure 10-18. *Checking the numerical attributes (Total Amount)*

Figure 10-19 is the result of the correlation matrix for the internal data; highly correlated variables are a disadvantage when using Euclidian distance. The pairwise table is given in Figure 10-20 (sorted in descending order). Based on this result, we will eliminate Item Count, No of C, Email Status = P, and Quantity as they have high correlation values.

Attribut...	Billing S...	Billing S...	Billing S...	Registr...	Registr...	Email St...	Email St...	No of C	No of D	No of N	Quantity	Item Co...	Total A...	No of D...	Discoun...	Unit Price	ProfitM...
Billing St...	1	-0.395	-0.512	0.139	-0.139	-0.115	0.115	0.173	0.034	0.347	-0.043	-0.060	-0.048	0.260	0.190	-0.085	0.109
Billing St...	-0.395	1	-0.484	-0.252	0.252	0.181	-0.181	-0.312	0.081	-0.343	-0.094	-0.060	-0	-0.124	0.209	0.008	0.093
Billing St...	-0.512	-0.484	1	0.085	-0.085	-0.047	0.047	0.103	-0.102	-0.035	0.122	0.109	0.045	-0.142	-0.360	0.074	-0.183
Registrat...	0.139	-0.252	0.085	1	-1	-0.255	0.255	0.162	-0.093	-0.079	0.206	0.181	-0.046	0.293	-0.045	0.300	-0.344
Registrat...	-0.139	0.252	-0.085	-1	1	0.255	-0.255	-0.162	0.093	0.079	-0.206	-0.181	0.046	-0.293	0.045	-0.300	0.344
Email St...	-0.115	0.181	-0.047	-0.255	0.255	1	-1	0.333	-0.296	0.052	0.034	0.311	0.153	-0.806	0.108	0.301	0.012
Email St...	0.115	-0.181	0.047	0.255	-0.255	-1	1	-0.333	0.296	-0.052	-0.034	-0.311	-0.153	0.806	-0.108	-0.301	-0.012
No of C	0.173	-0.312	0.103	0.162	-0.162	0.333	-0.333	1	-0.332	-0.078	0.803	0.826	0.527	-0.414	-0.008	0.523	-0.193
No of D	0.034	0.081	-0.102	-0.093	0.093	-0.296	0.296	-0.332	1	-0.204	-0.152	-0.430	-0.333	0.418	0.120	-0.310	0.377
No of N	0.347	-0.343	-0.035	-0.079	0.079	0.052	-0.052	-0.078	-0.204	1	-0.394	-0.282	0.152	-0.071	-0.171	-0.532	0.430
Quantity	-0.043	-0.094	0.122	0.206	-0.206	0.034	-0.034	0.803	-0.152	-0.394	1	0.813	0.455	0.107	-0.064	0.654	-0.219
Item Cou...	-0.060	-0.060	0.109	0.181	-0.181	0.311	-0.311	0.826	-0.430	-0.282	0.813	1	0.522	-0.357	-0.043	0.733	-0.351
Total Am...	-0.048	-0	0.045	-0.046	0.046	0.153	-0.153	0.527	-0.333	0.152	0.455	0.522	1	-0.326	-0.286	-0.009	-0.071
No of Da...	0.260	-0.124	-0.142	0.293	-0.293	-0.806	0.806	-0.414	0.418	-0.071	0.107	-0.357	-0.326	1	-0.036	-0.174	0.137
Discount...	0.190	0.209	-0.360	-0.045	0.045	0.108	-0.108	-0.008	0.120	-0.171	-0.064	-0.043	-0.286	-0.036	1	0.079	-0.043
Unit Price	-0.085	0.008	0.074	0.300	-0.300	0.301	-0.301	0.523	-0.310	-0.532	0.654	0.733	-0.009	-0.174	0.079	1	-0.518
ProfitMade	0.109	0.093	-0.183	-0.344	0.344	0.012	-0.012	-0.193	0.377	0.430	-0.219	-0.351	-0.071	0.137	-0.043	-0.518	1

Figure 10-19. *Correlation matrix*

First Att...	Second ...	Corr... ↓
No of C	Item Cou...	0.826
Quantity	Item Cou...	0.813
Email St...	No of Da...	0.806
Item Cou...	Unit Price	0.733
Quantity	Unit Price	0.654
Item Cou...	Total Am...	0.622
No of C	Quantity	0.603
No of C	Total Am...	0.527
No of C	Unit Price	0.523
Quantity	Total Am...	0.455
No of N	ProfitMade	0.430
No of D	No of Da...	0.418
No of D	ProfitMade	0.377
Billing St...	No of N	0.347
Registrat...	ProfitMade	0.344
Email St...	No of C	0.333
Email St...	Item Cou...	0.311

Figure 10-20. *Pairwise correlation*

To determine the number of k, we will use 2, 3, 4, and we select the best separated k. Import the data again and, this time, set the role of Customer ID to Id when importing so that we can link each customer to its cluster. The settings in Figure 10-21 will give the visualized cluster in Figures 10-22a (K=2), 10-22b (performance of K=2), 10-23a (K=3), 10-23b (performance of K=3), 10-24a (K=4), and 10-24b (performance of K=4).

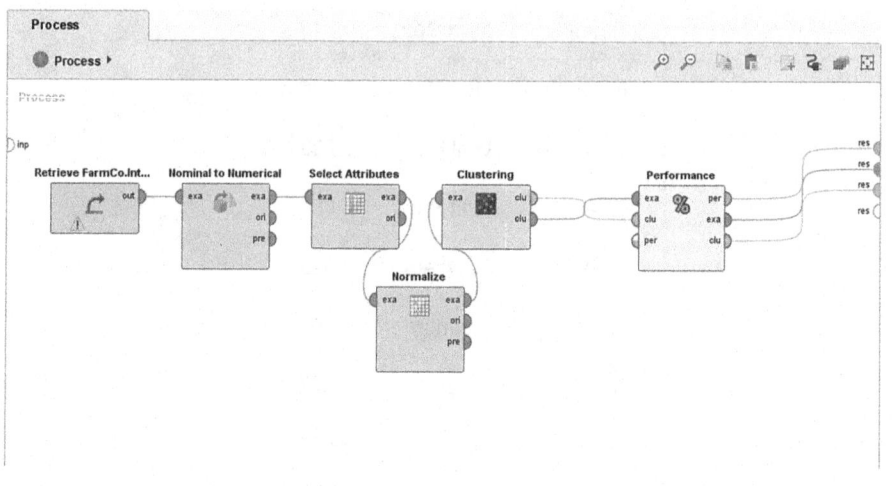

Figure 10-21. *The clustering process (k-means)*

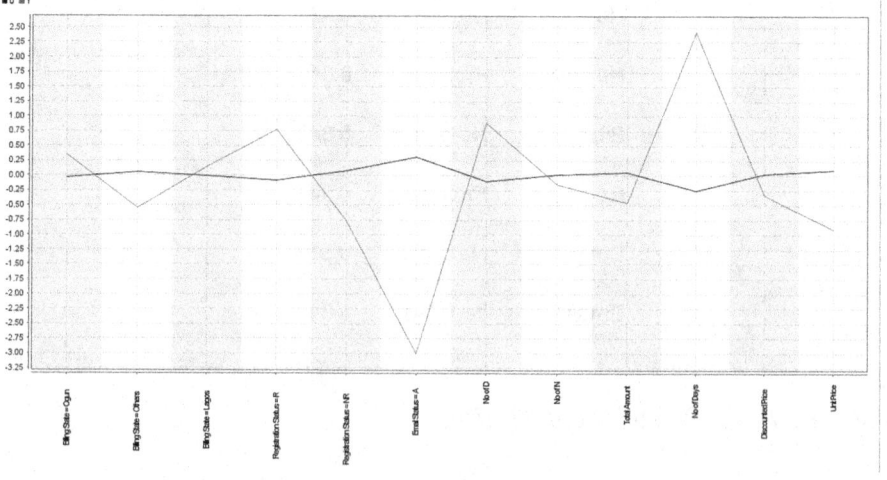

Figure 10-22a. *Cluster profile for K=2*

PerformanceVector

```
PerformanceVector:
Avg. within centroid distance: -9.482
Avg. within centroid distance_cluster_0: -10.059
Avg. within centroid distance_cluster_1: -3.997
Davies Bouldin: -1.071
```

Figure 10-22b. *Cluster performance of K=2*

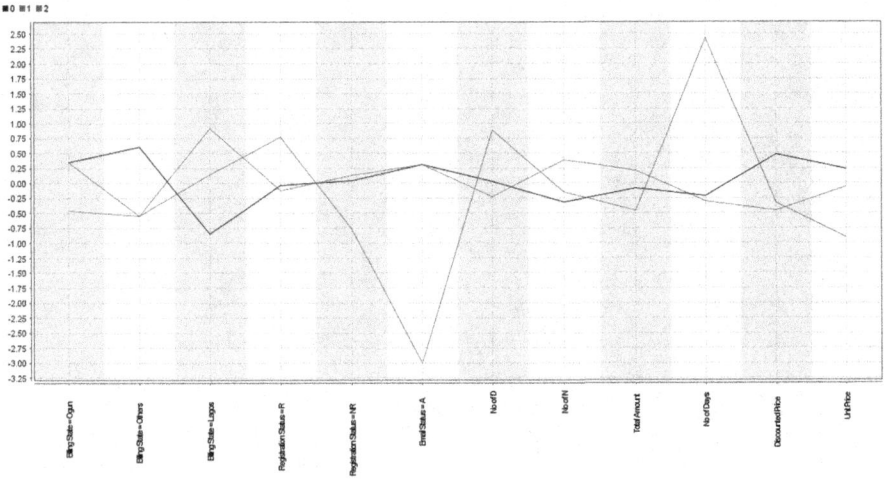

Figure 10-23a. *Cluster profile for K=3*

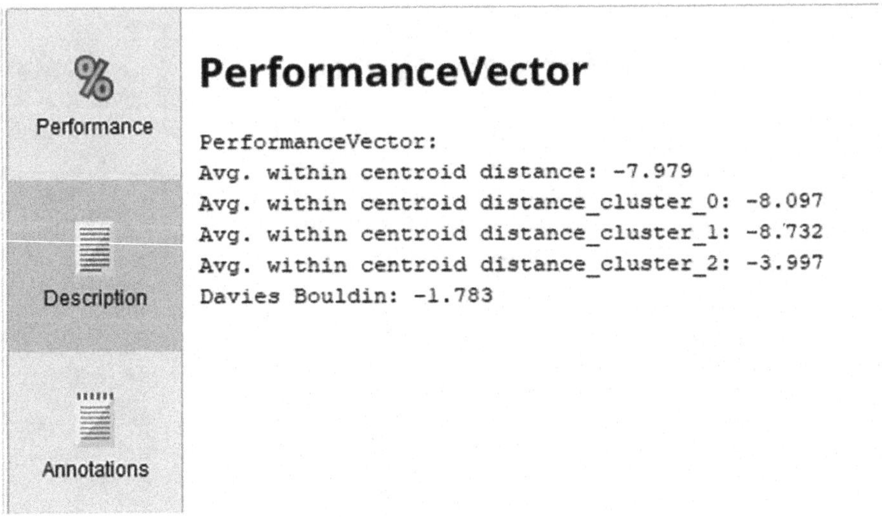
Performance
Description
Annotations
PerformanceVector
PerformanceVector:
Avg. within centroid distance: -7.979
Avg. within centroid distance_cluster_0: -8.097
Avg. within centroid distance_cluster_1: -8.732
Avg. within centroid distance_cluster_2: -3.997
Davies Bouldin: -1.783

***Figure 10-23b.** Performance of K=3*

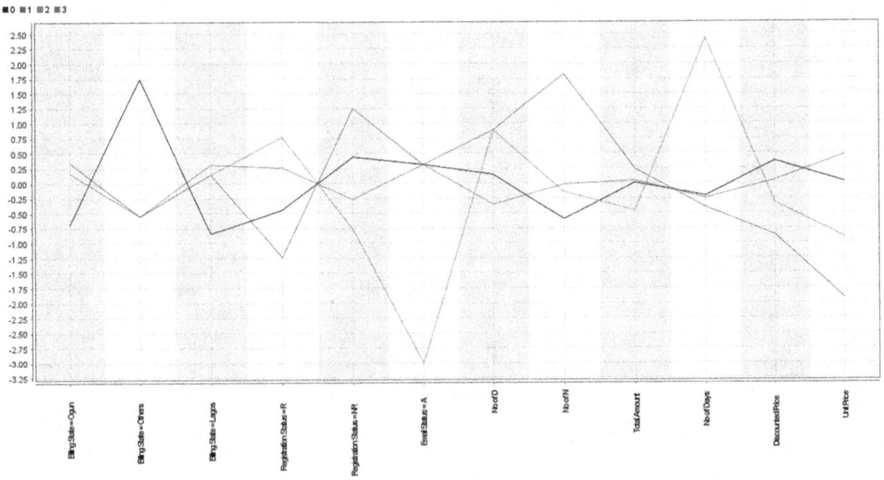
BillingState=Ogun
BillingState=Others
BillingState=Lagos
RegistrationStatus=R
RegistrationStatus=NR
EmailStatus=A
NoofD
NoofN
TotalAmount
NoofDays
DiscountedPrice
UnitPrice

***Figure 10-24a.** Cluster profile for K=4*

PerformanceVector

```
PerformanceVector:
Avg. within centroid distance: -6.618
Avg. within centroid distance_cluster_0: -6.391
Avg. within centroid distance_cluster_1: -7.464
Avg. within centroid distance_cluster_2: -3.997
Avg. within centroid distance_cluster_3: -7.009
Davies Bouldin: -1.404
```

Figure 10-24b. *Performance of K=4*

From the preceding results, we select K=3 due to the following reasons:

- avg._within_centroid_distance: The average within-cluster distance is calculated by averaging the distance between the centroid and all examples of a cluster. The smaller, the better.

- davies_bouldin: The algorithms that produce clusters with low intra-cluster distances (high intra-cluster similarity) and high inter-cluster distances (low inter-cluster similarity) will have a low Davies-Bouldin index; the clustering algorithm that produces a collection of clusters with the smallest Davies-Bouldin index is considered the best algorithm based on this criterion.

Having decided on K=3 for the internal data, to choose the best segmentation result for the problem at hand (internal data), we can use the profit made as a metric and then for each segment calculate the mean profit for all members in the cluster. There are other metrics that can be used such as cancel rates, etc., depending on the application scenario. The output of the cluster (Figure 10-25) can be linked with the original FarmCo. Internal.xlsx to know the profit of each cluster member and then calculate each cluster's mean profit made. This is revealed in Table 10-2. Notice that the profit made is not part of the attributes for clustering as it will not be in the external data.

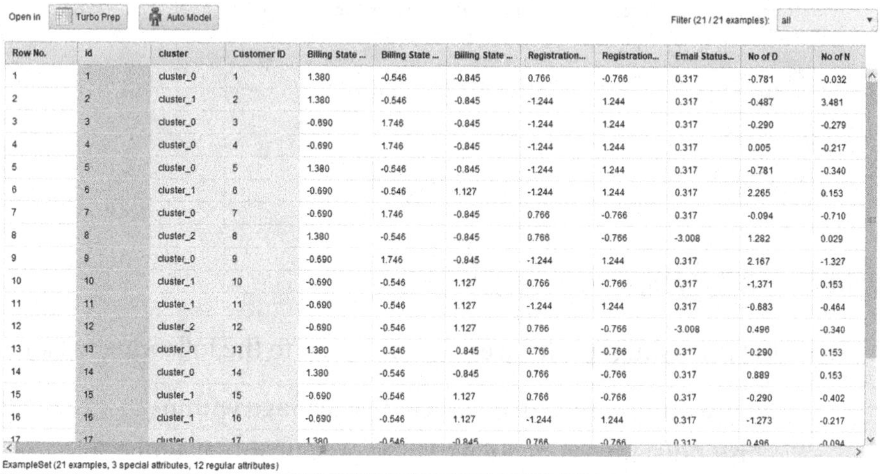

Open in Turbo Prep Auto Model Filter (21 / 21 examples): all

Row No.	id	cluster	Customer ID	Billing State ...	Billing State ...	Billing State ...	Registration...	Registration...	Email Status...	No of D	No of N
1	1	cluster_0	1	1.380	-0.546	-0.845	0.766	-0.766	0.317	-0.781	-0.032
2	2	cluster_1	2	1.380	-0.546	-0.845	-1.244	1.244	0.317	-0.487	3.481
3	3	cluster_0	3	-0.690	1.746	-0.845	-1.244	1.244	0.317	-0.290	-0.279
4	4	cluster_0	4	-0.690	1.746	-0.845	-1.244	1.244	0.317	0.005	-0.217
5	5	cluster_0	5	1.380	-0.546	-0.845	-1.244	1.244	0.317	-0.781	-0.340
6	6	cluster_1	6	-0.690	-0.546	1.127	-1.244	1.244	0.317	2.265	0.153
7	7	cluster_0	7	-0.690	1.746	-0.845	0.766	-0.766	0.317	-0.094	-0.710
8	8	cluster_2	8	1.380	-0.546	-0.845	0.766	-0.766	-3.008	1.282	0.029
9	9	cluster_0	9	-0.690	1.746	-0.845	-1.244	1.244	0.317	2.167	-1.327
10	10	cluster_1	10	-0.690	-0.546	1.127	0.766	-0.766	0.317	-1.371	0.153
11	11	cluster_1	11	-0.690	-0.546	1.127	-1.244	1.244	0.317	-0.683	-0.464
12	12	cluster_2	12	-0.690	-0.546	1.127	0.766	-0.766	-3.008	0.496	-0.340
13	13	cluster_0	13	1.380	-0.546	-0.845	0.766	-0.766	0.317	-0.290	0.153
14	14	cluster_0	14	1.380	-0.546	-0.845	0.766	-0.766	0.317	0.889	0.153
15	15	cluster_1	15	-0.690	-0.546	1.127	0.766	-0.766	0.317	-0.290	-0.402
16	16	cluster_1	16	-0.690	-0.546	1.127	-1.244	1.244	0.317	-1.273	-0.217
17	17	cluster_0	17	1.380	-0.546	-0.845	0.766	-0.766	0.317	0.496	-0.094

ExampleSet (21 examples, 3 special attributes, 12 regular attributes)

Figure 10-25. *Clustered dataset*

Table 10-2. *Mean Cluster Profits*

Cluster Name	Average Profit
Cluster_0	172,000
Cluster_1	745,555
Cluster_2	175,000

Next, we try to explain the characteristics of all the preceding clusters as follows, as obtained from Figure 10-23:

Cluster_0

Billing State=Lagos is slightly low.

Billing State Others is slightly high.

This cluster does not have striking attributes that can distinguish it.

Cluster_1

Billing State=Lagos is slightly high.

Other attribute characteristics are not striking.

Cluster_2

Email Status=A is very low.

No of Days is very high.

These two attributes mostly separate items on this cluster.

Since Cluster_1 has the highest mean profit, ideally we would want to use the characteristics of cluster 1 to look for a similar cluster in the external data and then target it (going by the option of the best-case scenario), because it has a high number of customers and also high mean profit made.

External data 5a–e

The next set of steps will help to segment the external customers.

Import the external data (FarmCo.External), and exclude the following while importing: Customer ID, Shipping State, Quantity, No of C, Item Count. This is because they are not used for internal data clustering.

After examining the external data (FarmCo.External.csv) for missing values, we have the statistics in Figure 10-26. There are missing values for some attributes. This will be dealt with by the *replace missing value* operator later.

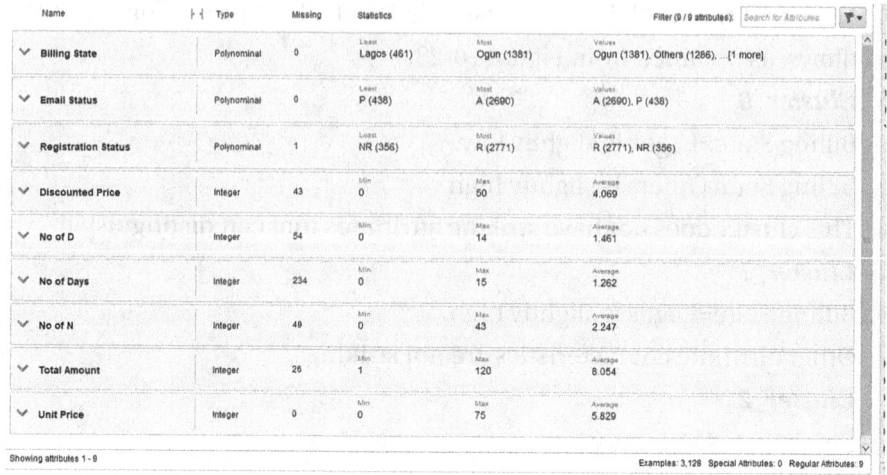

Figure 10-26. *Summary statistics*

Checking the numerical attributes for outliers (using scatter plots), we discover that all the attributes don't have outliers. Figure 10-27 is an example.

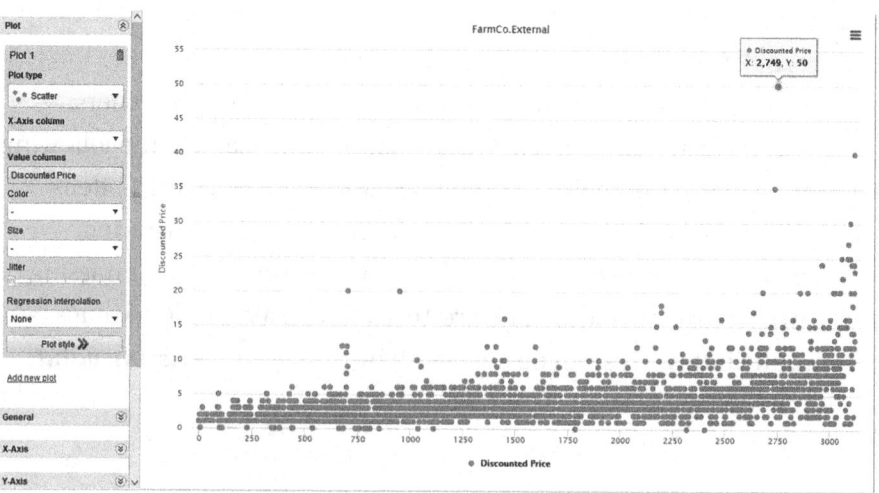

Figure 10-27. *Checking the numerical attributes (Discounted Price)*

We can check the correlation matrix for the external data (Figure 10-28). It is important to note that we have to use the same attributes used for the internal data segmentation for the external data segmentation to be able to link the two clustering outputs together.

Attribut...	Billing S...	Billing S...	Billing S...	Email St...	Email St...	Registr...	Registr...	Discoun...	No of D	No of D...	No of N	Total A...	Unit Price
Billing St..	1	-0.743	-0.370	-0.422	0.422	0.179	-0.179	0.416	0.404	0.437	0.377	0.364	0.494
Billing St..	-0.743	1	-0.347	0.313	-0.313	-0.158	0.159	-0.301	-0.383	-0.305	-0.339	-0.308	-0.394
Billing St..	-0.370	-0.347	1	0.157	-0.157	-0.032	0.030	-0.165	-0.035	-0.192	-0.058	-0.083	-0.145
Email St..	-0.422	0.313	0.157	1	-1	-0.145	0.145	-0.246	-0.205	-0.277	-0.157	-0.145	-0.278
Email St..	0.422	-0.313	-0.157	-1	1	0.145	-0.145	0.246	0.205	0.277	0.157	0.145	0.278
Registrat..	0.179	-0.158	-0.032	-0.145	0.145	1	-0.998	0.105	0.109	0.142	0.119	0.077	0.147
Registrat..	-0.179	0.159	0.030	0.145	-0.145	-0.998	1	-0.104	-0.108	-0.142	-0.119	-0.077	-0.148
Discount..	0.416	-0.301	-0.165	-0.246	0.246	0.105	-0.104	1	0.502	0.518	0.493	0.508	0.625
No of D	0.404	-0.383	-0.035	-0.205	0.205	0.109	-0.108	0.502	1	0.386	0.623	0.645	0.532
No of Da..	0.437	-0.305	-0.192	-0.277	0.277	0.142	-0.142	0.518	0.386	1	0.442	0.394	0.601
No of N	0.377	-0.339	-0.058	-0.157	0.157	0.119	-0.119	0.493	0.623	0.442	1	0.735	0.533
Total Am..	0.364	-0.308	-0.083	-0.145	0.145	0.077	-0.077	0.508	0.645	0.394	0.735	1	0.481
Unit Price	0.494	-0.394	-0.145	-0.278	0.278	0.147	-0.148	0.625	0.532	0.601	0.533	0.481	1

Figure 10-28. *Correlation matrix (external data)*

Import the external data again and, this time, set the role of Customer ID to Id. Exclude the following while importing: Customer ID, Shipping State, Quantity, No of C, Item Count. The settings in Figure 10-29 will give the visualized cluster in Figures 10-30a (K=2), 10-30b (performance of K=2), 10-31a (K=3), 10-31b (performance of K=3), 10-32a (K=4), and 10-32b (performance of K=4). Note that when selecting attributes, select the ones used for internal clustering.

315

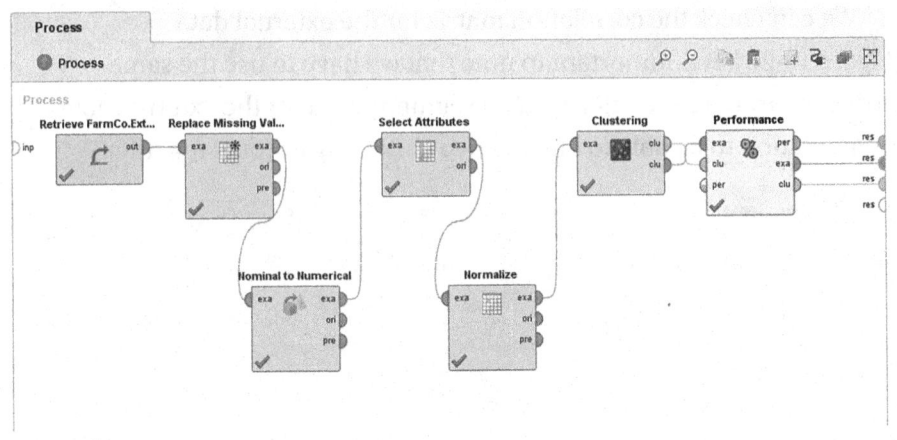

Figure 10-29. *Clustering the external data*

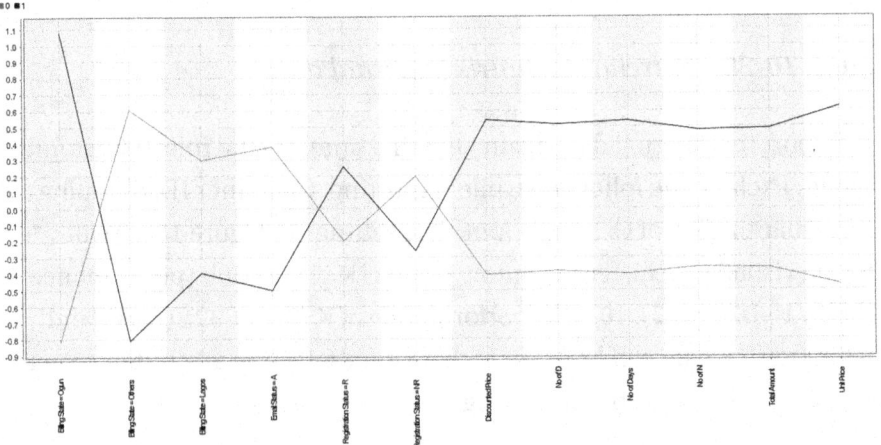

Figure 10-30a. *Cluster profile for K=2 (external data)*

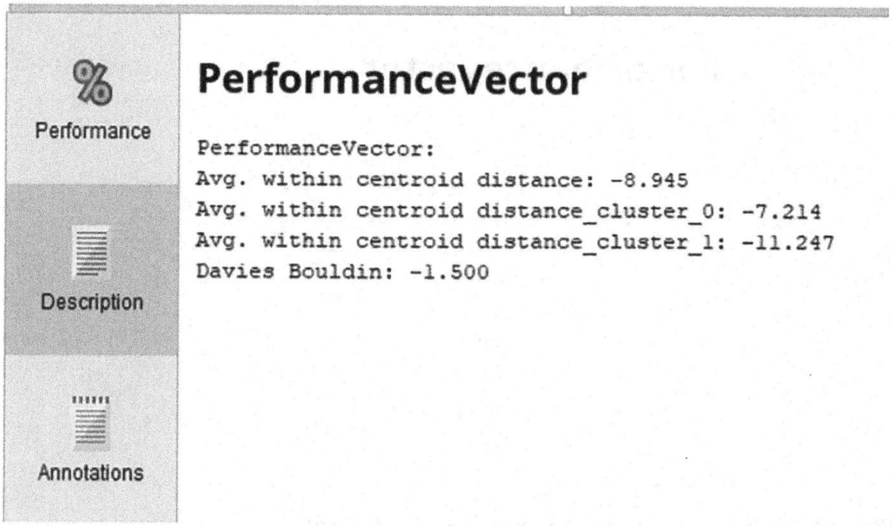

PerformanceVector

```
PerformanceVector:
Avg. within centroid distance: -8.945
Avg. within centroid distance_cluster_0: -7.214
Avg. within centroid distance_cluster_1: -11.247
Davies Bouldin: -1.500
```

Performance

Description

Annotations

Figure 10-30b. *Performance of K=2 (external data)*

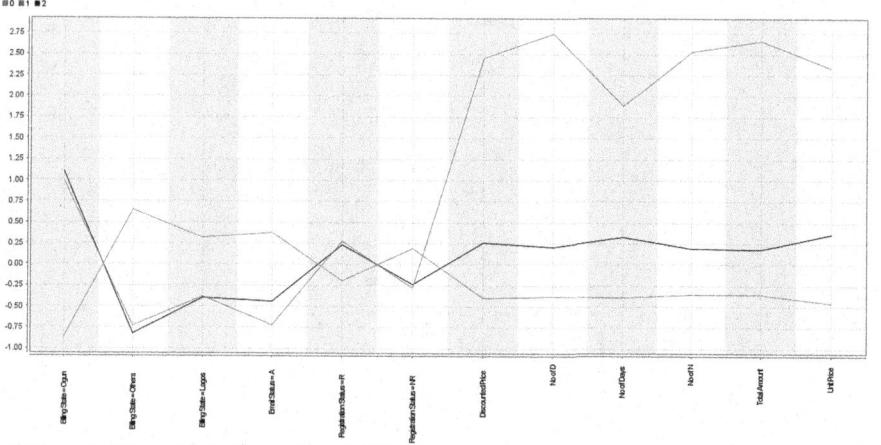

Figure 10-31a. *Cluster profile for K=3 (external data)*

Performance

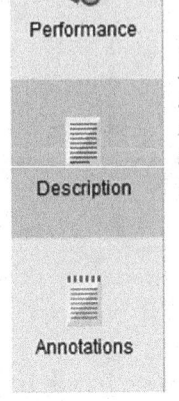

Description

Annotations

PerformanceVector

```
PerformanceVector:
Avg. within centroid distance: -7.704
Avg. within centroid distance_cluster_0: -7.129
Avg. within centroid distance_cluster_1: -30.060
Avg. within centroid distance_cluster_2: -5.785
Davies Bouldin: -1.423
```

Figure 10-31b. *Performance of K=3 (external data)*

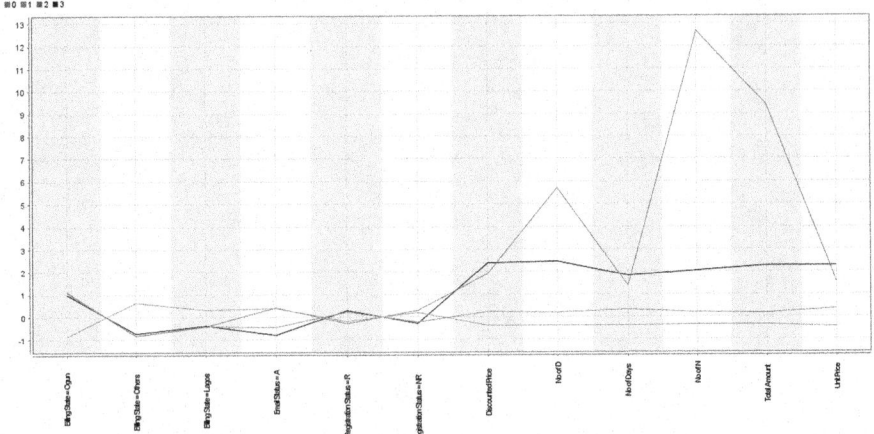

Figure 10-32a. *Cluster profile for K=4 (external data)*

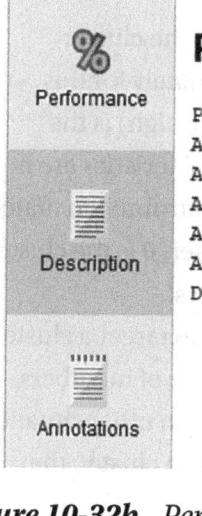

PerformanceVector

```
PerformanceVector:
Avg. within centroid distance: -7.435
Avg. within centroid distance_cluster_0: -5.661
Avg. within centroid distance_cluster_1: -7.133
Avg. within centroid distance_cluster_2: -41.792
Avg. within centroid distance_cluster_3: -22.577
Davies Bouldin: -1.276
```

Figure 10-32b. *Performance of K=4 (external data)*

From the preceding results, we select K=2 due to its smallest davies_bouldin value.

The characteristics of the clusters are given as follows:

Cluster_0

Billing State=Ogun is low.

Billing State=Others is slightly high.

Billing State=Lagos is slightly high.

Email Status=A is slightly high.

DiscountedPrice is low.

Cluster_1

Billing State=Ogun is very high.

Billing State=Others is low.

Billing State=Lagos is slightly low.

Email Status=A is low.

DiscountedPrice is slightly high.

(Note that other k values that give better separation and distinct characteristics can be tried out.)

Linking both segmentation processes

So after segmenting external data, we want to select the cluster that matches Cluster_1 from the internal data. Unfortunately for this demonstration, Cluster_1 (Billing State=Lagos is slightly high) is the only striking characteristic as the other attributes' characteristics are not striking. Regardless, we still proceed by using only this attribute to match the characteristics of the external data to find that Cluster_0 is the cluster with such characteristics along with other characteristics.

The recommendation therefore is that if we intend to target a cluster that has a potential of high profit made and large number of members, Cluster_0 in the external data is the recommended one. In other situations, where there are more distinguishing characteristics of the cluster, the recommendation would have been richer.

1. ***Display and present the result.***

   Based on the audience in question, these results need to be presented in simple charts supported with necessary visual diagrams to depict the best option and the reason for selecting such an option.

10.3 Problems

1. Using the data named PoultryNigeria.gephi, use Gephi to arrive at Figure 10-5.

2. Visualize the data named SampleHistorical.xlsx for classification using a side-by-side boxplot.

3. Repeat the RFM Segmentation (under the section *customer loyalty intervention*) in this chapter using the data RFMFoodAgric.xlsx and FinalRFM.xlsx. Are you able to arrive at the same results?

4. How would you present the result gotten from problem 3 to the stakeholders?

10.4 References

1. RFM Segmentation, www.optimove.com/resources/learning-center/rfm-segmentation

2. Anthony Capetola (February 28, 2021) This is How You Can Use Predictive Analytics to Sell Smarter Through Email, www.coredna.com/blogs/predictive-analytics

3. Jennifer Xue, How to Use Predictive Analytics for Better Marketing Performance, www.singlegrain.com/digital-marketing-analytics/how-to-use-predictive-analytics-for-better-marketing-performance/

4. Divya Chandana (May 2021) Exploring Customers Segmentation With RFM Analysis and K-Means Clustering, https://medium.com/web-mining-is688-spring-2021/exploring-customers-segmentation-with-rfm-analysis-and-k-means-clustering-118f9ffcd9f0

Data Files

Chapter 2

- FarmCo.External.xlsx
- OnlineQuestion1.xlsx
- OnlineQuestionAfterMissing.csv
- OnlineQuestionAfterOutlier.csv

Chapter 5

- SuperstoreNigeria.xls (original data source is `https://community.tableau.com/s/question/OD54TOOOOOCWeX8SAL/sample-superstore-sales-excelxls`)
- OnlineQuestion1.xlsx
- customersNigeria.xlsx
- chapter5Assign.xlsx

Chapter 6

- ToyotaCorollaData.xlsx (original source: `www.dataminingbook.com/book/r-edition`)
- ToyotaCorollaNewData.xlsx
- FoodAgricGroup.xlsx

A. I. Tolulope, *Data Science and Analytics for SMEs*, https://doi.org/10.1007/978-1-4842-8670-8

Chapter 7

- Bank.xlsx (original source: www.dataminingbook.com/book/r-edition)

- Churn_ModelingNew.csv (original source: www.dataminingbook.com/book/r-edition)

- Churn_Modelling.csv

- FramesData.xlsx

- InternalData.xlsx

- newBank.xlsx

- NoodleRetailLRExternal.xlsx

- NoodleRetailLRInternal.xlsx

- Prospects.xlsx

- TrainingDS3.csv

- TrainingDS3NewData.csv

Chapter 8

- Alegra.xlsx

- dynamic.gexf

- EdgeListProducts.csv

- edgeslistStaff.csv

- NodeListProducts.csv

- nodeslabelStaff.csv

- NoodleRetail.xlsx

- NoodleRetailNormalize.csv

- shoes.xlsx

- StatisticStaff.csv

- VegetableTransactions.csv

Chapter 10

- PoultryNigeria.gexf

- FoodAgricGroup.xlsx

- FoodAgricTransAsso.xlsx

- SampleHistorical.xlsx

- FinalRFM.xlsx

- RFMFoodAgric.xlsx

- FarmCo.External.xlsx

- FarmCo.Internal.xlsx

Index

L

M

N, O

Printed in the United States
by Baker & Taylor Publisher Services